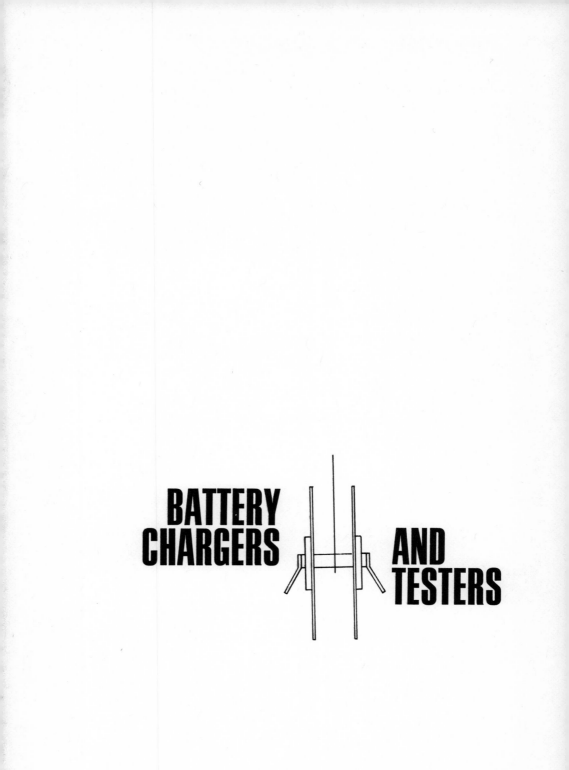

BATTERY CHARGERS

AND TESTERS

BATTERY CHARGERS AND TESTERS

OPERATION
REPAIR
MAINTENANCE

Charles R. Cantonwine, E.E.

CHILTON BOOK COMPANY
Philadelphia New York London

Published in Philadelphia by Chilton Book Company
and simultaneously in Ontario, Canada,
by Thomas Nelson & Sons, Ltd.

ISBN 0-8019-5621-8
Library of Congress Catalog Card Number 73-153135
Designed by Cypher Associates, Inc.
Manufactured in the United States of America

PREFACE

Electric batteries are used on more equipment, and in increasing numbers every year. Most of these batteries are the rechargeable type, and need battery chargers and battery testers.

Battery charges and testers wear out from normal use, and often fail because of abuse. Therefore, they require service from time to time. Up to a few years ago, battery chargers were relatively simple and could be serviced by anyone having only a little knowledge of electricity. Today, however, they are more complicated, with automatic solid state electronics being used in automatic voltage regulators, alternator protectors, and built-in battery testers.

To publish the manufacturer's wiring diagrams and parts lists of every battery charger and tester model manufactured under many brand names would require several thick volumes and would not give the complete technical data needed to service this equipment. The wiring diagrams furnished by the manufacturers are usually in "block" form, showing only the external connections to the various assemblies. This is because they prefer to have an entire assembly replaced rather than to have incompetent technicians attempt internal repairs to the more complicated control panels, such as those for alternator protectors, voltage regulators, timers, and the like. This practice will probably continue until the repair industry can prove that it is capable of properly testing and repairing these units, and until proper test instruments and repair equipment are made available. Better availability of all parts, and more technical information will also help the cause.

Very little has been published on the application of the various electrical and electronic components used in battery charger circuits. I believe that this book provides the basic knowledge of all the electronic components used, and fills the gap between common knowledge and its specific application to battery chargers and battery testers.

An appendix listing sources of supply for battery charger and tester parts and equipment should be helpful when there is a need for catalog information.

Although this book is written primarily as a service manual, it should be of interest as a general reference to anyone concerned with the maintenance of batteries and battery service equipment, such as salesmen, servicemen, mechanics, technicians, engineers and manufacturers of associated equipment.

All of the circuits used in this manual were obtained by tracing them from chargers and testers actually serviced by the author over a period of many years and are up to date.

Charles R. Cantonwine E.E.

CONTENTS

Preface

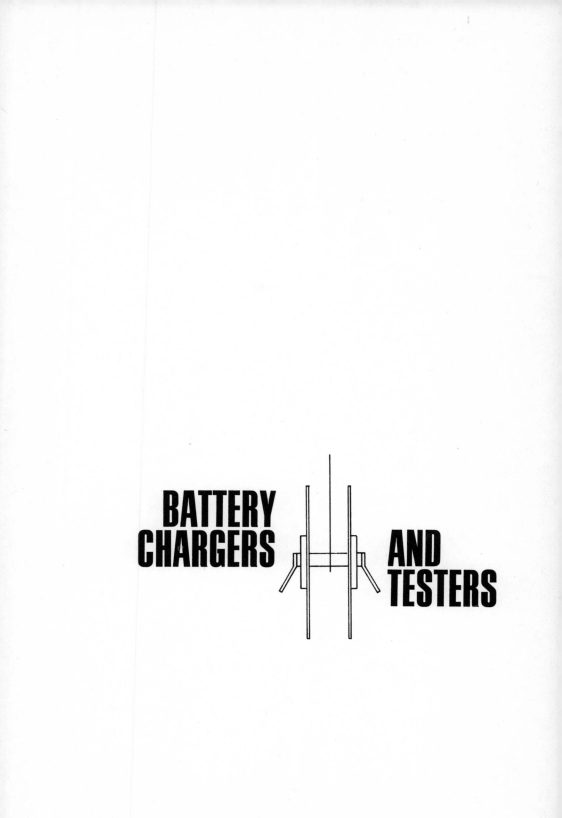

BATTERY CHARGERS AND TESTERS

SECTION I

1
CHARGING
CIRCUITS

A battery charger is a device for renewing the electrical charge in a storage battery, or other rechargeable battery, by applying a direct current of the proper voltage and current to the battery to be charged.

Simple Battery Charging Circuit From Direct Current

If a direct current source having a voltage higher than the battery BAT in Figure 1-1 is available, it is necessary to use only a current limiting

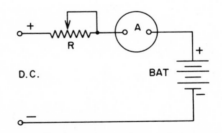

Figure 1-1   Simple battery charging circuit from a direct current source

resistor R to control the rate of charge, and an ammeter A to read the rate of charge. Resistor R may be a fixed value or adjustable, to vary the charge rate. Direct current is usually not readily available, but alternating current is generally available at 115 volts, 230 volts and higher, as a source of power.

Simple Battery Charging Circuit From Alternating Current

Alternating current must be changed to direct current to charge a battery. This is done by using an A.C. electric motor to drive a D.C. generator of the proper voltage, or it is done by using a rectifier circuit with or without a transformer.

Before the development of the modern selenium and silicon type rectifiers, several kinds of rectifiers were used with various degrees of success.

a. The mercury vapor rectifier which is still used in industrial applications.

b. Electrolytic rectifiers using two different metals in a solution, such as aluminum and steel in a solution of ammonium phosphate.

c. Mechanical rectifiers driven by a synchronous motor, or a tuned vibrator switching arrangement.

d. The tungar type of bulb with a filament enclosed in a glass bulb is still widely used.

e. Copper oxide and copper sulphide rectifiers are obsolete for battery charger use.

This manual covers the applications of the modern solid state rectifier systems and tungar bulbs.

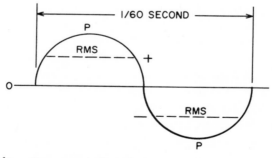

Figure 1–2   Alternating current sine wave

A modern rectifier is a solid state device, with no moving parts, that has low resistance (passes a high current) when the polarity of the current is in one direction, but has a very high resistance (passes very little or no current) when the polarity of the current is in the opposite direction.

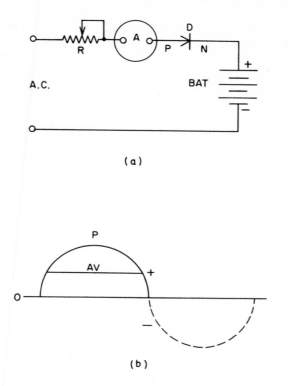

(b)

Figure 1–3 Simple battery charging circuit from an AC source

Alternating current reverses its polarity from zero to positive, back to zero, to negative and back to zero 60 times a second (or 60Hz), as shown in Figure 1-2. The highest point of the sine wave P is called the peak voltage, and the dotted line represents the RMS, or root mean square, voltage which is .886 or 88.6% of the peak voltage. This RMS voltage is the value read on an A.C. voltmeter; 0 represents zero voltage.

A rectifier connected as shown in Figure 1-3 will block the negative half of the cycle and allow only the positive half of the cycle to pass to the battery. In Figure 1-3a, R is a current limiting resistor and may be adjustable as shown, A is an ammeter to read the D.C. charging current, D is a half-wave rectifier or diode having an anode P and a cathode N, with current flow from the anode to the cathode to the battery BAT.

Figure 1-3b shows how the rectifier allows the positive half of the sine wave to pass, shown in solid lines, and blocks the negative half of the sine wave, as shown in dotted lines; the line marked 0 is zero current; the line marked AV is the average positive D.C. current that passes, which is .707 or 70.7% of the peak current P, and is the value read on the D.C. ammeter. This is known as a half-wave rectifier because it passes only half of the full sine wave, and is connected for positive output.

Using the circuit of Figure 1-3a, the power loss in the resistor R is an important factor for A.C. voltages that are much higher than the battery voltage, and is therefore limited to use on 115 volts A.C. for very low currents, and where the cost and space are factors, such as on small portable appliance batteries. By reversing the connection to the rectifier, a negative output is obtained, making it necessary to reverse the connections to the battery and ammeter, also. With this reversed connection, the rectifier passes only the negative half of the sine wave and blocks the positive half.

Rectifier Types

Rectifiers are made in many different styles, shapes, sizes, and of different materials, but they all serve the same purpose, some better than others. The following three common types are used in modern battery chargers.

A. SELENIUM RECTIFIERS. The selenium rectifier cell is made in the shape of a disc or plate consisting of four essential layers as shown in Figure 1-4a. The aluminum base plate A, which is always the anode A, is a strong mechanical support for the other layers, and dissipates the heat developed; the thin selenium layer F is applied to the base plate A; an extremely thin barrier or blocking layer E is applied next, which is where the rectifier action takes place; and counter-electrode layer C is applied last. This is made of a metal alloy, and is always the cathode C. The alloy layer C is recognized by the silvery crystalline appearance, if unpainted. This is not selenium, as is commonly thought, but it is a thin metallic contact plate. It does not extend to the edge of the base plate A, but is slightly smaller than blocking layer E to eliminate the possibility of a direct short to the aluminum anode at the edges. The area of this alloy cathode is the effective rectifying area of the cell. Some rectifiers are painted with a good heat conducting finish, which also retards corrosion.

Some brands of selenium cells have a higher conductivity, and operate at higher temperatures than others. Some have a higher peak reverse voltage rating than others. None of these characteristics can be determined by the eye, so it is important to replace selenium rectifiers with original factory parts, or parts of known equality. Selenium cells are constantly being improved.

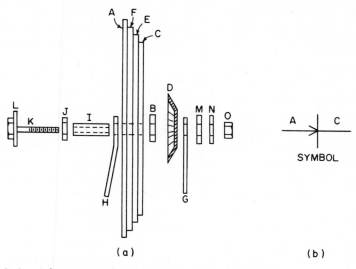

(a)   (b)

Figure 1–4   Selenium rectifier cell construction and symbol

In Figure 1-4a the cathode terminal strip G is in contact with the metal alloy coating C through a solid conducting washer, or a spring contact plate D with a pressure limiting, precision insulating washer B. The anode terminal strip H contacts the uncoated side of the base plate A; tubing I is an insulator tube; washer J is an insulating washer; bolt or rivet K is a clamping and mounting stud; piece L is a metal washer; piece M is an insulating washer; piece N is a metal washer; nut O and bolt K clamp the entire assembly together. Often, the assembly is clamped by a rivet and washers instead of the bolt and nut.

Each selenium cell or plate, as used in battery chargers, will usually stand 50 PIV. However, higher voltage cells are available. Two or more cells can be connected in series to increase the voltage rating.

The average voltage drop across a selenium cell is approximately 1.5 volts.

Selenium rectifiers for low current chargers are usually mounted on the inside wall or back panel, which acts as a heat sink. For higher current chargers, the selenium plates act as fins, and are located in an air stream from a motor driven fan.

Selenium rectifiers are also found in other shapes, such as round or tubular, in the smaller ratings.

The schematic symbol shown in Figure 1-4b indicates the anode A which corresponds to terminal strip H, and cathode C which corresponds to terminal strip G of Figure 1-4a.

B. SILICON DIODES. Silicon diodes are much smaller than selenium rectifiers of the same rating. They consist of a small pellet of silicon enclosed in a metal housing, except the very small ones. This metal housing may be the type that is pressed into a hole in the heat sink, or it may be stud mounted, being held in place on the heat sink by a nut. Smaller diodes are mounted by two leads and usually have no heat sink except that provided by the conductivity of the two leads to the connecting terminals.

Whereas, the selenium cell anode is always the base plate, the silicon diode can be manufactured so the metal case is either the anode or the cathode. The polarity on the press-in type is indicated by color code. Red lettering or red paint indicates a positive output base, which is considered normal polarity. The terminal is the anode and the base is the cathode. Black lettering or black paint indicates a negative base, or reverse polarity. The terminal is the cathode and the base is the anode. On the stud mounted type, the polarity is usually indicated by an arrow, which points from the anode to the cathode.

A single silicon pellet will usually stand about 50 PIV, but two or more pellets connected in series within the single housing provide higher PIV ratings, such as 100, 200, 300, 400.

The approximate voltage drop across the average 50 PIV silicon diode is about 0.8 volts, and increases proportionally with an increase in the PIV rating.

Rectifiers having the lowest voltage drop are the most efficient. The power loss in watts, and hence the heat generated, is a product of the voltage drop multiplied by the current. That is why silicon diodes are so widely used, the power loss is low, and they can stand a high temperature.

The current rating of a rectifier is determined entirely by its temperature limitation. The operating temperature is determined by the power loss in the rectifier, and how fast this heat can be dissipated. Diodes, except the very small ones, are always mounted on a heat sink. The heat sink is either fastened to the case, as an additional heat sink, and may be cooled by convection, or by a stream of air from a motor driven fan.

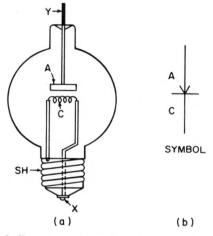

SYMBOL

(a)   (b)

Figure 1–5   Tungar bulb construction and symbol

c. BULB TYPE RECTIFIER. The bulb type rectifier, or tungar bulb is shown in Figure 1-5a. It consists of a heated filament C as the cathode and a metallic plate A connected to terminal Y as the anode, enclosed in a gas filled glass bulb. The filament voltage is approximately 2-2.5 volts, and draws between 15 and 20 amperes. The filament voltage is applied through the shell SH and contact X of the mogul screw base. The heated filament C ionizes the gas in the bulb to make it conduct electricity when the anode A is positive, but block the current flow when the anode A is negative. The schematic symbol is shown in Figure 1-5b. Tungar bulbs are available in ampere ratings of 2, 5, 6, and 15. However, the 2 and 5 ampere bulbs have a slightly different appearance than shown in Figure 1-5a, but they operate the same.

The voltage drop across the tungar type bulb rectifier is about 10 volts. The bulb is usually air cooled by natural air circulation through the vents in the charger case.

The shell SH is always connected to the positive charger output because it can carry heavier current than the contact X and must carry both the filament current and the charging current.

Tungar bulbs should be replaced when the filament is sagging. This indicates at least 1500 hours of normal service, or severe overloading. A sagging filament causes a higher voltage drop and increases the power loss, requiring that the rate switch be turned up higher.

Also, there is available a silicon diode type rectifier with a screw base and a finned aluminum heat sink that is interchangeable with the tungar type bulb rectifier. It uses a press-in type silicon diode, which can be economically replaced with a 25 ampere 300-400 PIV silicon diode having a positive base. A similar unit can be made up in the shop using readily available materials by following the construction details in section V.

Types of Transformers

The simple charger circuit shown in Figure 1-3 is practical only for low currents of a few milliamperes. For higher currents, a transformer is used that reduces the line voltage from 120 or 240 to a voltage only slightly higher than the D.C. voltage of the battery being charged. The amperage in the secondary is stepped up in proportion. For example, a 120 volt primary drawing 10 amperes at full load could put out approximately 120 amperes from a 10 volt secondary.

Transformers used in battery chargers may have either an auto-transformer type of winding, which has only one winding, or an isolated type of winding having a primary winding insulated from a separate secondary winding. Either type of winding may be found on either a "core" type or a "shell" type of magnetic laminated iron core.

A. AUTO-TRANSFORMER TYPE WINDING. The auto-transformer type has only one winding wound on the core, which serves both as a primary winding and as a reduced voltage secondary winding tapped as shown in Figure 1-6. The transformer T has the primary P connected to 115 V.A.C. at terminals B and W. The hot or black wire is connected to B, and the ground or white wire is connected to W. This keeps the secondary close to ground potential to avoid shock. The secondary S voltage appears at terminals H and G, and is tapped off at taps A, B, C, and D by tap switch SW.

The wire sizes on the windings must be larger on the secondary S section than on the primary P section because the secondary section must carry the stepped up secondary current as well as the primary current. The auto-transformer is the most efficient and least expensive to construct, but has the disadvantage of shock hazard. This type must be permanently connected to the line and grounded as shown or, for portable use, a polarized plug should be used.

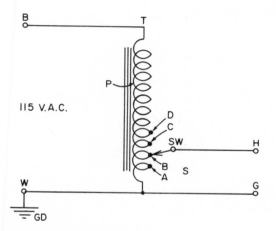

Figure 1–6   Auto-transformer type winding

The auto-transformer is usually used for low current chargers of 6-15 amperes for multiple charging of as many as 6 or more 12 volt batteries in series, and may use a half-wave, full-wave, or bridge rectifier circuit. Tungar bulbs and silicon diodes are commonly used in half-wave or full-wave rectifier circuits using auto-transformers.

B. ISOLATED WINDING TYPE TRANSFORMER. The isolated winding type transformer is shown in Figure 1-7 where transformer T uses a separate primary winding P which is insulated from a separate secondary winding S, but electro-magnetically coupled together. The 115 V.A.C. is applied to the primary winding P at terminals B and W, and the secondary S voltage appears across terminals H and G.

The voltage ratio between the primary and secondary windings is, as in the auto-transformer, proportional to the number of turns of wire in each winding. For example, if the 115 volt primary has 100 turns of wire and the secondary has 10 turns of wire, the secondary voltage (no load) would be 11.5 volts. This is a turns ratio of 10 to 1. If the secondary has a wire size 10 times the area of the primary wire, it would put out 10 times the current drawn by the primary winding. In actual design, a few more turns are added to the secondary to compensate for the voltage drop and leakage reactance under load. This secondary voltage, under load, is quite critical in battery charger transformers unless a wide range of adjustment taps is provided.

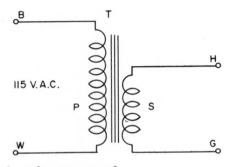

Figure 1–7   Isolated winding type transformer

c. "CORE" TYPE TRANSFORMER CORE. The "core" type transformer core structure is shown in Figure 1-8. Figure 1-8a shows the auto-transformer winding wound on core X, where P is the primary connection and A, B, C, and D are the secondary taps. Figure 1-8b shows the isolation type transformer windings wound on core X, where P is the primary connection and S is the secondary connection.

For battery charger service, both the primary and secondary must be wound equally on both legs of the core to prevent saturation of the iron core by the D.C. current flowing through the secondary winding. For this reason, the core type is not used for auto-transformers except for full-wave rectifier circuits where the winding can be wound in two equal sections, one-half on each side of center-tap CT.

d. "SHELL" TYPE TRANSFORMER CORE. The "shell" type core structure

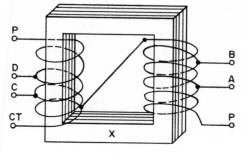

(a)

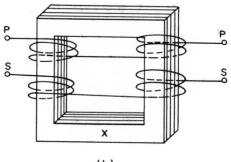

(b)

Figure 1–8   "Core" type transformer core

is shown in Figure 1-9. Figure 1-9a shows the auto-transformer type wind-ing where P represents the primary winding connection. Figure 1-9b shows the isolation type winding where P represents the primary winding connection. In both figures, the secondary winding connection is shown as S.

In both the core and shell types, the primary windings and secondary windings may be wound on top of each other, either one being wound on the bottom first, or they may be alternated side by side in a "pie" arrangement.

If rewinding of battery charger transformers is undertaken, the original winding must be duplicated in every detail, because the manufacturer had in mind certain characteristics for maximum performance. Some trans-

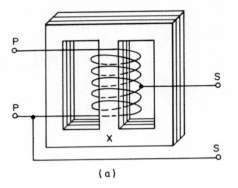

(a)

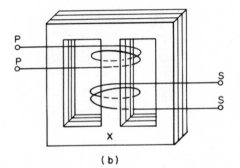

(b)

Figure 1–9   "Shell" type transformer core

formers cannot be rewound due to their welded or sealed construction.

Transformers used in battery chargers require special considerations not required for resistance loads because of the D.C. current in the secondary winding.

Half-Wave Battery Charging Circuit

A simple, transformer type, battery charger is shown in Figure 1-10 using one rectifier plate or diode D in a half-wave circuit. The same action takes place as illustrated in Figure 1-3. This circuit is usually limited to slow or trickle chargers in the low current range. The transformer T steps the 115 V.A.C. line voltage down to slightly above the battery voltage. For trickle chargers, the peak voltage only needs to be above the battery voltage, but for heavier current chargers, the RMS voltage must be higher than the battery voltage. The maximum D.C. current must not exceed the rating of diode D. Usually, the diode D is chosen to operate at only about ½ of its maximum capacity, as a safety factor.

For a half-wave rectifier circuit, the transformer should be rated to have a volt-ampere capacity of at least 3½ to 4½ times the D.C. watts output due to saturation, which cannot be eliminated.

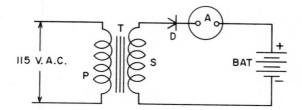

Figure 1–10   Half-wave battery charging circuit

Full-Wave, Center-Tap, Charging Circuit

A full-wave rectifier, using a center-tapped transformer secondary, is shown in Figure 1-11a. Here, the secondary of the transformer T has a third lead brought out from the exact electrical center of the winding (center tap), which goes to the negative side of the battery. Rectifiers D1 and D2 are the same, and are connected for positive output. A is a D.C. ammeter and BAT is the battery.

The rectifying action of a full-wave rectifier is shown in Figure 1-11b. During the first half of the A.C. cycle rectifier D1 conducts because it is positive, and rectifier D2 blocks, because it is negative. During the second half of the A.C. cycle, the polarity of the transformer secondary winding reverses so that rectifier D2 is now positive and conducting, and rectifier D1 is negative and blocking. Therefore, we have two positive pulses of D.C. current going to the battery during one cycle. Twice as much current is supplied to the battery with a full-wave rectifier as with a half-wave rectifier.

By reversing the connections of the rectifiers, ammeter and battery, the D.C. output would be negative, and the transformer center tap would be positive.

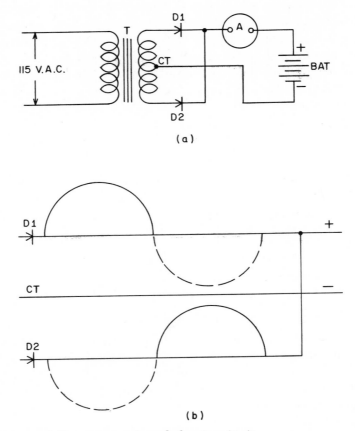

Figure 1–11   Full-wave, center-tapped charging circuit

The total D.C. current output of the full-wave rectifier is equal to the combined amperage rating of the two rectifiers, or twice the rating of a single rectifier.

For a center-tapped transformer secondary and full-wave rectifier, the volt-ampere rating of the transformer must be equal to 1½ to 2 times the D.C. watts output. If a core type transformer is used, it must be built so that one half of the secondary winding is wound on each leg and connected in series. Otherwise, the transformer will be saturated as it is in half-wave circuits. Shell type construction does not have this problem.

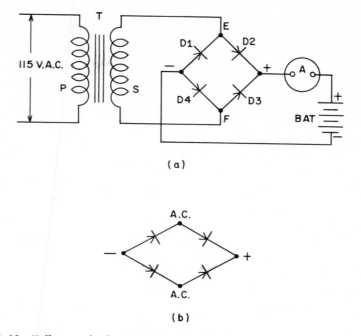

Figure 1–12   Full-wave, bridge type charging circuit

Only one half of the total D.C. current goes through each rectifier and each half of the secondary winding.

Full-Wave, Bridge Charging Circuit

A full-wave bridge rectifier circuit is shown in Figure 1-12a. No center-tap is required with the bridge circuit, since the secondary carries the full D.C. output current. The total D.C. amperage output is only one half of the combined output rating of the 4 rectifiers D1, D2, D3 and D4, and is twice the output of a single rectifier rating. Here, at the instant when the transformer T secondary S is positive at point E, it is negative at point F. Rectifier D2 conducts a positive current to the positive D.C. output, and rectifier D1 blocks the flow of current. Likewise, with point F negative, rectifier D4 conducts the negative current to the negative D.C. output, and rectifier D3 blocks the flow of current. As the polarity of the transformer secondary reverses, point E now becomes negative, while point F is positive. When point E is negative, rectifier D1 now conducts a negative current to the negative D.C. output, and rectifier D2 blocks the flow of current. With point F positive, rectifier D3 conducts a positive current to positive D.C. output, and rectifier D4 blocks the flow of current. The rectified output has the same wave form as shown in Figure 1-11b.

It is easy to remember the bridge circuit if the schematic is simplified as shown in Figure 1-12b. The arrows of the rectifiers all point in the same direction, that is, from negative to positive.

For the full-wave bridge rectifier circuit, the volt-ampere rating of the transformer should be 1⅓ to 1¾ times the D.C. watts output.

The schematic wiring diagrams of Figures 1-10, 1-11 and 1-12, show the three basic, single-phase rectifier circuits used in battery chargers. These circuits are for single battery voltage, with no rate of charge adjustment, giving maximum starting charge on a dead battery and tapering off to a lower charging rate as the battery charge builds up.

Three-Phase, Half-Wave, Charging Circuit

Although most battery chargers encountered will be for single-phase operation, there are some industrial chargers that operate from a three-phase power supply, and you should know the three-phase rectifier connections.

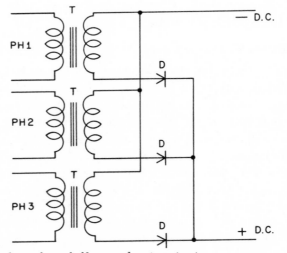

Figure 1–13   Three-phase, half-wave charging circuit

The three-phase, half-wave rectifier circuit is shown in Figure 1-13, using three separate, identical, single-phase transformers T. All rectifiers D are the same and are connected for positive output. PH1, PH2 and PH3 represent the three phases. The total output is equal to three times the rating of each rectifier D.

When using a three-phase transformer with the half-wave circuit, it should be connected in a certain way to avoid the D.C. saturation that occurs when using three separate single-phase transformers. The primary should be connected in a Delta connection. The secondary is divided into six parts, with one set of three coils (one from each phase) connected in Wye. The output of each leg of the Wye passes through another half-phase placed on another leg of the transformer.

Three-Phase, Full-Wave Center Tap, Charging Circuit

A three-phase, full-wave center-tapped circuit is shown in Figure 1-14, that may use a single three-phase transformer or three separate single-

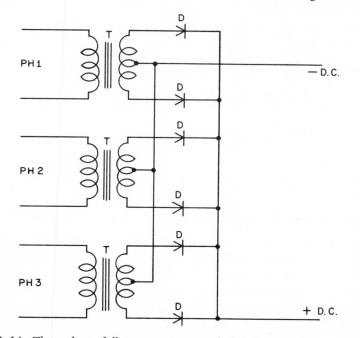

Figure 1–14 Three-phase, full-wave center-tapped charging circuit

phase transformers. Rectifiers D are shown connected for positive output. PH1, PH2 and PH3 show the three phases of the power supply.

The total output equals six times the rating of each rectifier D.

Three-Phase Bridge Charging Circuit

A three-phase bridge circuit is shown in Figure 1-15, where PH1, PH2 and PH3 represent the three-phase power supply, which can be the secondaries of three separate single-phase transformers, a single three-phase transformer secondary, or the output of a three-phase alternator, such as used on vehicles. The total D.C. output equals three times the rating of each rectifier D.

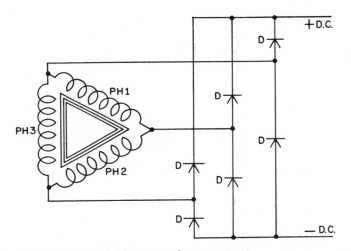

Figure 1–15 Three-phase, bridge type charging circuit

Constant-Current, Resonant, Leakage-Reactance Transformer

The most common type of transformer used in battery charger circuits is the constant-voltage type, that is, the output of the charger is maintained at a practically constant voltage, giving a high initial charging current to a discharged battery and tapering off to a lower charging current as the battery voltage builds up.

There is another type, known as a constant-current, resonant, or leakage-reactance transformer, which has many uses in battery charger service.

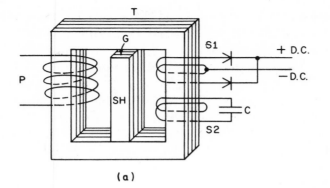

(a)

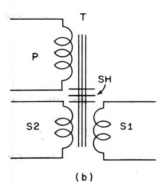

(b)

Figure 1–16 Constant-current, resonant, leakage-reactance transformer

To charge Edison type cells, consisting of a nickel-iron structure and a caustic solution, a charger capable of delivering a constant charging current over the entire charging cycle is necessary.

There are automatic multiple battery chargers that will charge any size battery or any combination of batteries in series, consisting of three to fifty-four, two volt cells, without switching, using a constant-current charger supplying about 6 amperes at all battery voltages, or even when the clips are shorted together.

These constant-current chargers may use a special transformer as shown in Figure 1-16. The transformer T is usually of the core type, but it has a primary P on one leg of the core, and may have one or more secondaries S1 and S2 on the opposite leg. Secondary S1 is a conventional secondary supplying the rectifiers and D.C. output. Secondary S2 is a high voltage winding connected to an oil-type, continuous-duty condenser C. There is a magnetic shunt SH placed between the primary and secondary windings with an air gap G. This arrangement is known as a "leakage-reactance" transformer. With no load on the secondary S1, most of the flux created by the primary P will reach secondaries S1 and S2, and will have full no-load voltage. When a load is placed on secondary S1, part of the primary flux is by-passed, or shunted, through the air gap G and magnetic shunt SH. The voltage across secondary S1 will drop rapidly about in proportion to the load. The current supplied by secondary S1 will be almost constant at various loadings. To make this current more constant regardless of load, the secondary S2 and condenser C (having capacitive reactance) must be the correct value to be equal and opposite to the combined inductive reactances of the primary P, secondary S2, and the leakage-reactance of the transformer T. At resonance, the current in the secondary S2 and condenser C circuit and the voltage across condenser C are both at a maximum, just as in any elementary tuned or resonant circuit. The resonant frequency is 60 Hz for a 60 Hz power supply.

The actual value of condenser C, in most cases, should have a mfd value within 1% of the original factory condenser, which usually is marked on the case in ink, in addition to the nominal value which is usually stamped on the case. For example, for a 20 mfd original, the replacement must be not more than 20.2 mfd or less than 19.8 mfd. The voltage across condenser C may be as high as 300-400 volts.

The presence of a separate secondary S2 and a condenser C does not always mean that the charger is of the constant-current type. A modified form of this arrangement is sometimes used to correct the power factor, and to maintain the same charging rate, initial and taper, over a wide range of line voltage variations. In these chargers, the value of the condenser C is not so critical. A condenser with nominal value can be used as a replacement.

The leakage-reactance transformer schematic symbol is shown in Figure 1-16b, and its construction is shown in Figure 1-16a. The symbol shows, in the manufacturer's schematic, that the transformer has additional leakage-reactance over and above the small leakage-reactance present in constant-voltage transformers. However, the schematic will not indicate the degree of leakage-reactance, or whether the circuit is resonant, or of the constant-current type. By studying the panel lay-out and the instructions, this can usually be determined.

If the condenser C is open or disconnected, the D.C. output will drop considerably in either type. Some transformers have a thermal overload device built into the transformer, and connected in series with the secondary S2 to automatically reduce the charging rate if the transformer becomes overheated. Full charge is restored when the transformer cools off.

Dual-Voltage, Single-Rate Charger

Some battery chargers will charge a single-voltage battery at a single charging rate, and others will charge a single-voltage battery at a high and a low rate, or intermediate rates. Others are more versatile, and will charge batteries of different voltages, and at various rates. Any of these may have different arrangements of switches, circuit breakers, meters, fuses, timers or other automatic controls, and use various rectifier arrangements.

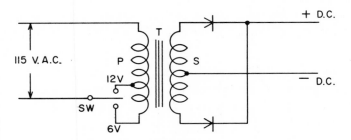

Figure 1–17   Dual-voltage, single-rate charger

Selection of various voltages and charging rates is done by switching to different taps on the transformer primary winding, secondary winding, or both, or by switching the rectifier or secondary connections.

The circuits shown in Figures 1-10, 1-11 and 1-12 show simple single-voltage, single-rate chargers using the three basic rectifier circuits: half-wave, full-wave center-tap, and full-wave bridge.

The remainder of this chapter is devoted to all the combinations of charging circuits found in battery chargers.

A simple, dual-voltage, 6 and 12 volt charger is shown in Figure 1-17, where the primary P of transformer T has two taps, one for 6 volts and the other for 12 volts. Switching is done by a SPDT switch SW, which may be a rotary, toggle, slide or rocker type switch, and usually has an "off" position. This circuit would be the same for a single-voltage charger having a high and low charging rate. The high rate would be on the 12V connection and the low rate on the 6V connection. However, in this case, there would be fewer turns of wire between the 6V and 12V connections. The secondary S may be found connected to any one of the various types of rectifiers and rectifier circuits.

It is very important, when replacing the rectifiers in this circuit or in any other single-rate charger, to use the same type and size of rectifier having the same conductivity and voltage drop as the original. When the primary winding of the transformer has several taps to adjust the rate or if a ballast resistor is used, the rectifier replacement is not as critical.

Multiple-Voltage, or Multiple-Rate Charger

The circuit shown in Figure 1-18 has a transformer T, which may have any number of taps on the primary P, but six taps are typical, as shown here. This may be for a single battery voltage, with a wide range of charging rates, or it may be, for example, a 6/8/12 volt charger. In this

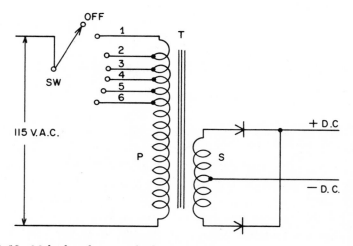

Figure 1–18 Multiple-voltage, multiple-rate charger

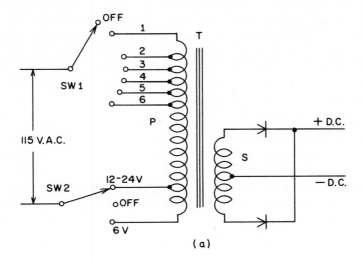

(a)

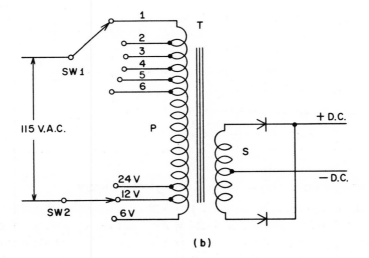

(b)

Figure 1–19 A 6/12/24 volt multiple-rate charger

case, selector switch SW could be for 6 volts using taps 1, 2 and 3 for low, medium and high respectively. For 8 volts, taps 2, 3 and 4 would be used for low, medium and high. For 12 volts, taps 4, 5 and 6 would be used for low, medium and high.

The rate switch SW may have a "dead" terminal for the "off" position, or it may have a separate line switch, or timer, to turn the A.C. on and off.

A 6/12/24 Volt, Multiple-Rate Charger

A battery charger capable of charging 6/12/24 volt batteries is shown in Figures 1-19a and 1-19b. It is similar to the charger in Figure 1-18 except for the additional taps on the lower end of the primary P of transformer T, and the additional switch SW2.

In Figure 1-19a, there are six rate selections for 6 volts, using switch SW1, but only three rates (taps 1, 2 and 3) for 12 volts, and three rates (taps 4, 5, and 6) for 24 volts. In Figure 1-19b, there are six rate selections (using switch SW1) on each setting of switch SW2 for voltages of 6, 12 or 24.

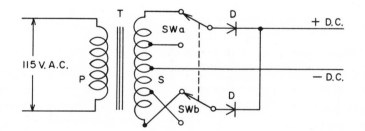

Figure 1–20   Dual-voltage charger, with a symmetrically-tapped secondary winding

Dual-Voltage, Symmetrical-Tapped Secondary Charger

The circuit shown in Figure 1-20 is for a fixed rate dual-voltage charger, such as for 6 or 12 volts, where transformer T has the secondary S tapped and connected as shown. Switch SW (consisting of SWa and SWb ganged together) is a DPDT switch having no "off" position, that selects for 6 or 12 volt battery. In the upper position, SW connects the two outer secondary terminals to the two rectifiers D, for a positive output of 12 volts. The center-tap of the transformer secondary is permanently connected to the negative output. In the lower position, switch SW connects for 6 volt charging.

This circuit is usually limited to small chargers of under 10 amperes, because of the larger size switch SW required for heavier currents.

Switch SW does not have an "off" position because it would only open the secondary circuit and not disconnect the A.C. line. This is usually done by pulling the A.C. plug, or by using a separate switch in the primary circuit.

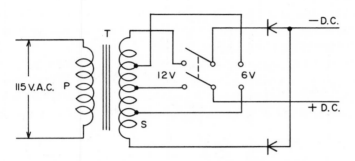

Figure 1–21   Dual-voltage charger with an unsymmetrically-tapped secondary winding

Dual-Voltage, Unsymmetrical-Tapped Secondary Charger

The circuit shown in Figure 1-21 is similar to the circuit of Figure 1-20, except that the center-tap connection for 6 and 12 volts is not common, as in Figure 1-20. Also Figure 1-21 has the rectifiers connected for negative output.

Dual-Voltage Charger, Series-Parallel Secondaries

A dual-voltage, 6/12 volt charger arrangement is shown in Figure 1-22, where transformer T has a primary P, which may be tapped for rate adjustments or untapped for a single rate charger, and two identical secondaries S1 and S2. Secondary S1 is center-tapped and provides 6 volts full-wave rectification through rectifiers D1 and D2. Secondary S2, provides 6 volts full-wave rectification through rectifiers D3 and D4. Switch SW connects the two 6-volt outputs in parallel for 6-volt batteries at twice the amperage output of a single section, and connects the two

6-volt outputs in series for 12-volt batteries at the same amperage output of a single section, but at twice the voltage.

One disadvantage of this circuit is that switch SW must be able to carry the heavy currents in the D.C. output, making it large and costly and subject to frequent failure if it is not conservatively rated. This circuit does have an advantage in that the rectifiers need only be one

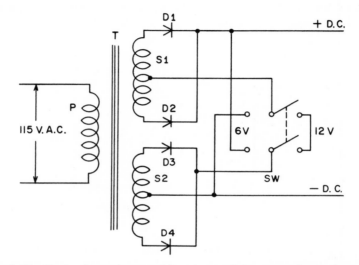

Figure 1–22 Dual-voltage charger with series-parallel connected secondary windings

half of the rating required for other circuits having the same output rating. For example, on a charger rated at 100 amperes at 6 volts and 50 amperes at 12 volts, the transformer must supply 600 watts output, which would limit the current to 50 amperes at 12 volts. Therefore, the rectifiers on 12 volts would be operating at one half of their 100 ampere capacity. With the transformer as the limiting factor, the rectifiers will carry 100 amperes at either 6 or 12 volts.

Some chargers are rated at the same charging current for both 6 and 12 volts, in which case both the transformer and rectifiers are operating at full capacity on the 12V connection; but on the 6V connection, only the rectifiers are operating at full capacity while the transformer is operating at only one half capacity.

In the circuit of Figure 1-22, the transformer and rectifiers are all operating at full capacity on either the 6 or 12 volt connection to provide twice the amperage on 6 volts as on 12 volts.

Dual-Voltage Charger, Bridge Circuit With Center-Tap

A dual-voltage charger for 6 or 12 volts, shown in Figure 1-23, provides either a full-wave bridge rectifier circuit for the higher 12-volt battery, or a full-wave, center-tapped rectifier circuit for the lower 6-volt battery. The transformer T is usually tapped in the primary P for rate adjustment by switch SW1 and the secondary S is center-tapped.

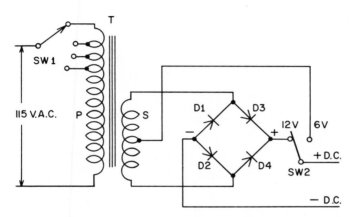

Figure 1–23   Dual-voltage charger with a bridge circuit and a center-tapped secondary winding

For charging 12 volt batteries, switch SW2 connects the positive output of the bridge circuit to the positive battery cable. The negative output of the bridge circuit is permanently connected to the negative battery cable. Rectifiers D1, D2, D3 and D4 form a bridge circuit. The center-tap is disconnected.

For charging 6 volt batteries, switch SW2 connects the center-tap of the transformer secondary to the positive output cable. The negative output cable is permanently connected at the junction of rectifiers D1 and D2, forming a full-wave center-tap connection. Rectifiers D3 and D4 are idle.

Since rectifiers D3 and D4 are used only on the lower current, higher voltage connection, they have only one half the rating of the rectifiers D1 and D2. The rectifiers D3 and D4 will frequently be smaller than rectifiers D1 and D2 in a selenium stack, or one pair may be silicon diodes and the other pair selenium rectifiers. Since switch SW2 must have the capacity to carry the full charging current, it must be a heavy duty switch.

If a replacement switch is not available, an emergency repair for switch SW2 can be made by bringing out the leads that end at the switch, through the front panel using strain relief bushings. Connect a large battery clamp to the positive D.C. cable, and terminate the ends of the positive 12 volt and positive 6 volt leads with a copper terminal. The battery clamp on the positive D.C. cable can then be clamped to either the 6 volt or 12 volt cable to select the correct battery voltage.

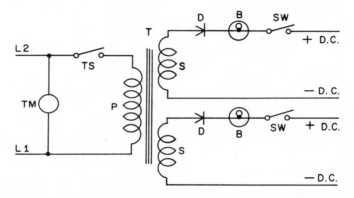

Figure 1–24 Typical multiple, isolated battery maintainer, source 7 Model 1159

The circuit of Figure 1-23 is useful to convert an old 6-volt only charger, if it has a center-tapped secondary and full-wave rectifiers, to a dual-voltage 6/12 volt charger by adding rectifiers D3 and D4 and switch SW2, particularly if the transformer primary is tapped to provide rate adjustment. The 12 volt amperes should be one half of the 6 volt current rating to keep within the power limits of the transformer.

Multiple Isolated Battery Maintainer

A battery charger used to maintain two or more batteries on stand-by service is shown in Figure 1-24. The transformer T has two or more secondaries S, each having a rectifier D, a lamp bulb B, and a switch SW. The bulb B shows if the circuit is charging, and also helps stabilize the half-wave rectified output. The primary P is supplied by 115 V.A.C. at L1 and L2 through timer switch TS. The electric timer motor TM operates continuously, driving a cam that opens and closes timer switch TS at predetermined intervals.

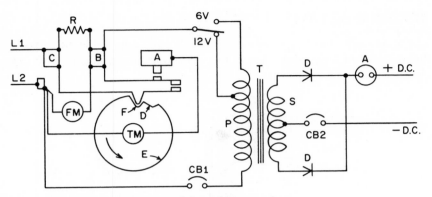

Figure 1–25   Typical source 4 Model R-60 battery charger

Model R-60 Charger

Figure 1-25 shows a two-rate, dual-voltage charger, with a timer that drops to a slow charge rate at the end of the timing cycle. When the timer is set for a timed high rate charge, contacts A, B and C are connected together, and the timer motor TM, fan motor FM, and the charger, all operate at full line voltage. When the timer reaches zero time, the contact arm C drops into the cam notch F, opening all contacts. This causes the voltage to the transformer and fan motor to be reduced to a slow charge through the 13 ohm resistor R, and the fan motor to run at reduced speed. However, the timer clock motor TM is disconnected and stops the rotation of the timer cam E. This slow charge will continue until the charger is disconnected by pulling the plug. By turning the timer knob to the "hold" position, the cam is turned so that the cam step D closes contacts B and C only, but does not close contact A. Therefore, the charger will operate continuously at high charge, and the fan motor will run at full speed, because the timer motor TM is disconnected and cannot turn the cam E.

On this single-rate charger, having no rate adjustment, it is important to use rectifiers that match the original in conductivity. These are of relatively low conductivity. The high conductivity plates will charge at a rate too high for the transformer to carry continuously.

Circuit breaker CB1 is held in contact with one of the selenium rectifier plates to disconnect the A.C. if the plate gets overheated, due to insufficient air circulation.

Models R-80 and R-100

Figure 1-26 shows a dual-voltage, 6/12 volt, adjustable rate charger, having a timer and a slow charge rate. When the timer is set for charging

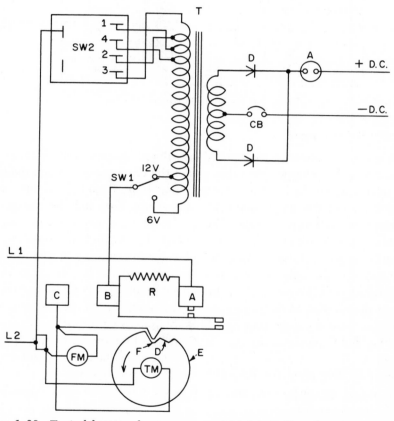

Figure 1–26 Typical battery chargers, source 4 Models R-80 and R-100

time, all contacts A, B and C are closed, and full line voltage is applied to the charger, timer motor TM, and fan FM. When the timer reaches the "slow" position, contacts A, B and C are opened as contact C arm drops into notch F of cam E. The slow-charge resistor R reduces the voltage across the charger transformer, and the timer disconnects the fan motor and timer clock motor. The charger continues to slow-charge until the A.C. plug is pulled. There is another position on the dial, adjacent to the "slow" position, marked "do not use". This position would cause cam step D to close contacts C and B, placing reduced voltage across the fan motor and timer motor. This operates the fan motor at reduced speed, and also may start the clock, rotating the cam, and starting the timing cycle again at full time and full charge. If the fan motor cannot start due to dry bearings, it could burn out due to insufficient air circulation. Some model R-80 and R-100 chargers use contacts A and B only as an "on-off" timer, and some model R-60 chargers have the "do not use" position. Switch SW1 is the voltage selector switch, and SW2 is the rate change switch. Rectifiers D are single-plate selenium rectifiers.

Model R-100-24

Figure 1-27 shows a 6/12/24 volt charger having an adjustable rate, and a timer with a "hold" position. When the timer is turned to charging time, contacts A, B and C close and full line voltage is applied to the charger transformer, T, fan motor FM, and timer clock motor TM. When the timer reaches zero time, all contacts open and the A.C. is disconnected. If the timer is moved to the "hold" position, the cam step D closes contacts B and C only, placing full line voltage across the charger transformer and the fan motor. Since contact A is open, the timer clock motor does not run, and the charger will operate at full charge until turned off or disconnected. This circuit uses a bridge rectifier circuit, consisting of the four selenium rectifier plates D. SW2 is a battery voltage selector switch, and SW1 is a rate adjustment switch.

This timer connection is also used on model S100, a 6/12 volt charger.

Ballast Primary, and Timer Circuit

Figure 1-28 shows a battery charger for 6/8/12/18 volts, without using a battery size selector switch. The ballast resistor R1 limits the charging current to a safe value regardless of the battery voltage between the

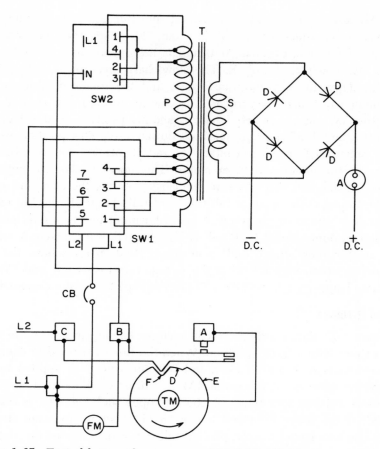

Figure 1–27 Typical battery charger, source 4 Model R-100-24

limits of 6-18 volts. This simplifies the operation, but it is less efficient due to the power lost in the ballast resistor. When the timer is turned on, the cam is in the position shown, with movable contact F riding on step 1a of the cam, and touching contact G. At the same time, movable contact C is riding on step 1 of the cam, touching contacts A and B. This applies full line voltage to the charger, timing motor TM and fan motor FM, while contacts A and B are closed, shorting out the trickle, or slow charge, resistor R2. As the timer motor moves the cam in the direction of the arrow, it reaches the point for the slow charge to start, when movable contact C drops into step 2 on the cam. This opens contact A, but not contact B. The short is removed from across the slow charge resistor R2, and the charging rate drops to a slow charge. The timer motor TM continues to run for a few minutes. Then movable contact F drops into step 3a on the cam and opens contact G. This lights the charge indicator bulb B1, which is now in series with the timer motor TM between contact B and L2. The bulb B1 is a neon bulb and does not draw enough current to operate the timer motor TM. Note that the

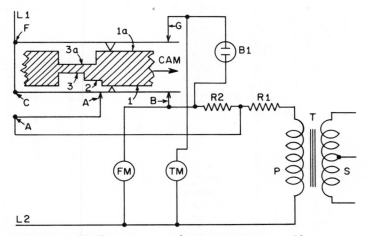

Figure 1–28 Typical ballast primary and timer circuit, source 12

charge indicator bulb does not light immediately when the timer starts the slow charge. The charger stays on slow charge, and the fan motor FM runs at full speed, until turned to "off" manually. When turned to the "off" position, movable contact C drops into step 3 on the cam, disconnecting contact B. All contacts are open and the charger is inoperative.

Only the primary circuit is shown because the secondary is conventional.

Another charger for 6/12/24 volts uses the same primary circuit, but uses a secondary circuit as shown in Figure 1-23. The low voltage position of the secondary selector switch would be for 6 and 12 volts, and the high voltage position of the switch would be for 24 volts. The ballast resistor R1 holds the charging current within limits.

Dual-Timer, Dual-Output Circuit Charger Model D36

Figure 1-29 shows a charger having two timers: timer motor TM1 having timer contacts TC1, and timer motor TM2 having timer contacts

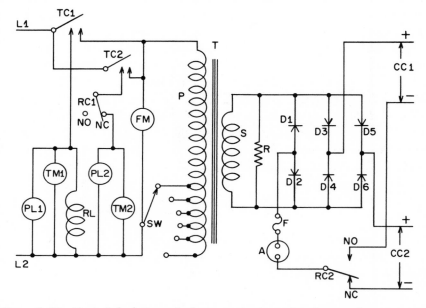

Figure 1–29 Typical dual-timer, dual-output circuit used in battery charger, source 12 Model D-36

TC2. Timer number 1 controls the time of the charge on charging circuit CC1, and timer number 2 controls the time of charging circuit CC2. A relay coil RL actuates two sets of contacts RC1 and RC2. The relay contacts are NC when the relay coil RL is not energized, but the NO contacts close, and the NC contacts open when the relay coil RL is energized.

To begin charging, both timers are set for the desired time, but only timer number 1 will operate, because relay coil RL is energized, and relay contacts RC1 and RC2 are in the NO position. This connects the

D.C. output to charging circuit CC1, and disconnects timer motor TM2 and pilot light PL2. When timer number 1 shuts off, the relay coil RL is de-energized so the relay contacts RC1 and RC2 close again switching the D.C. output to charging circuit CC2, starting timer motor TM2 and lighting pilot light PL2. When timer number 2 reaches the end of the timing period, it shuts the charger off.

The rectifier circuit is an unusual bridge circuit in that diodes D1 and D2 supply the negative polarity for both charging circuits CC1 and CC2 through relay contacts RC2, where the positive output diodes D3 and D4 supply positive polarity only to charging circuit CC1, and the positive output diodes D5 and D6 supply the positive polarity only to the charging circuit CC2. This makes it possible to charge two isolated, or interconnected, batteries together because the relay contacts RC2 keep the negatives separated, and the diodes D3, D4, D5 and D6 keep the positives isolated. SW is a rate control switch; R is a 100 ohm, 25 watt fixed resistor; fuse F protects the rectifiers; ammeter A indicates the charging current and FM is a fan motor for cooling the transformer and rectifiers. All diodes D1 to D6 are the same rating.

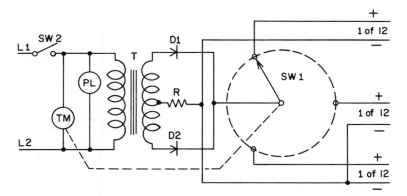

Figure 1–30   Typical automatic battery stock maintainer, source 12 Model 136

Automatic Battery Stock Maintainer Model 136

Figure 1-30 shows a battery maintainer, for charging as many as 36 batteries, having a 7 day timer motor TM driving a 36 contact rotary switch SW1. Each battery is charged once during the 7 day period. Only 3 circuits are shown, but there are actually 36 positive battery connections to the rotary switch SW1, and 36 common negative battery connections to the center-tap of transformer T, through a resistor R. Switch SW2 is a SPST "off" and "on" switch; PL is a pilot light. This charger is used to maintain idle batteries in stock.

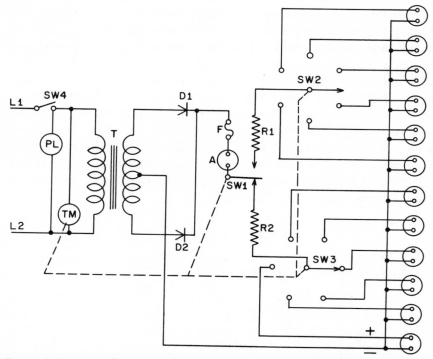

Figure 1–31   Typical automatic battery stock maintainer, source 12 Model 471

Automatic Battery Stock Maintainer Model 471

Figure 1-31 shows a battery maintainer using a time switch assembly consisting of a timing motor TM that drives two separate rotary switches SW2 and SW3, and a micro-switch SW1, which periodically alternates between the two switches SW2 and SW3, feeding the two battery banks. The twelve outlets to the batteries are shown at the right. SW4 is an "off" and "on" SPST switch; PL is a pilot light; T is the step-down transformer; D1 and D2 are the diodes; F is a 2 ampere fuse; A is the ammeter; and R1 and R2 are ballast resistors.

Dual-Voltage Battery Maintainer Model 613M

Figure 1-32 shows a battery maintainer using a thermal relay TR as a timer. TR is a plug-in type thermal relay, or flasher, with the numbers indicating the pin numbers. Switch SW1 is a DPDT switch, with an "off" position. When switch SW1 is in the upper "charge" C position, the charger operates continuously, but when SW1 is in the lower "maintain"

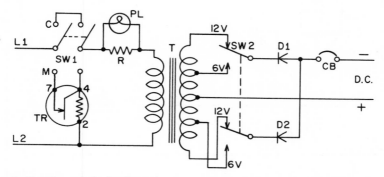

Figure 1–32   Typical dual-voltage battery maintainer, source 12 Model 613M

M position, the thermal element opens and closes contacts 4 and 7 of TR in a ratio of 36 to 1 (open 36 minutes, and closed one minute). R is a voltage-dropping resistor for low voltage pilot light PL; SW2 is a DPDT battery size selector switch; CB is a 6 ampere circuit breaker; D1 and D2 are diodes connected for negative output; and T is the step-down transformer.

Dual-Voltage Charger With Alternator Protector, Model 7660

Figure 1-33b shows the primary section, and Figure 1-33a shows the secondary section of a charger using two primary windings P1 and P2 connected in parallel for 12 volt operation. It uses winding P2 in series with part of the second primary winding P1 for 6 volt operation. The switching is done by a DPDT toggle switch SW1. This arrangement is often used to obtain the same ampere output on 6 volts as on 12 volts.

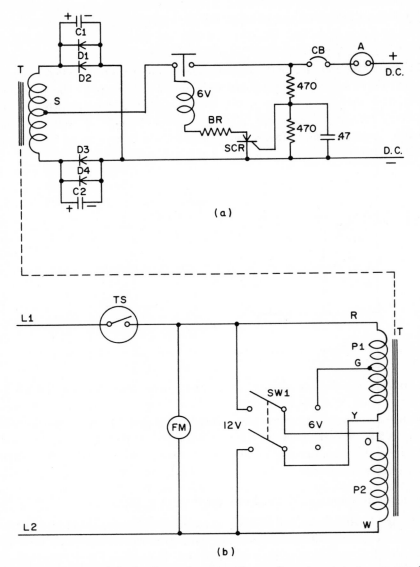

Figure 1–33 Typical dual-voltage charger with an alternator protector, source 13 Model 7660

FM is the cooling fan motor; TS is a mechanical spring clock time switch; capacitors C1 and C2 are 16 mfd, 50V electrolytic capacitors for transient voltage protection. D1, D2, D3 and D4 are identical, automotive-type diodes having a negative base, or output; CB is a 60A circuit breaker; and A is a 60 Ampere ammeter. The alternator protector circuit is on the right in Figure 1-33a and is explained in Chapter 16, Section III.

Automatic Golf-Cart Charger Models 400 (24V) and 500 (36V)

The automatic charger shown in Figure 1-34 is typical of a 24 or 36 volt charger used on golf-carts and small industrial trucks.

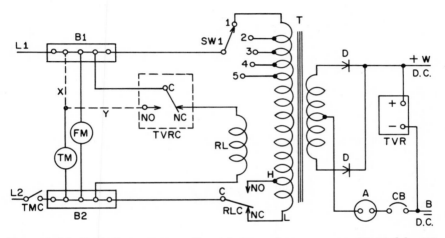

Figure 1–34   Typical automatic golf-cart battery charger, source 19 Models 400 (24V) and 500 (36V)

The selector switch SW1 is set for the initial charging rate, and the timer dial is turned to "on", closing timer contacts TMC. This starts the charger at full charging rate. Relay, or contactor, RL is energized through the NC terminal of the terminal voltage relay contacts TVRC, which in turn opens the NC contact and closes the NO contact of relay contact RLC, connecting the timer TM and fan motor FM to the high H tap on the transformer. As the battery builds up to operate the TVR, it opens the TVRC contact NC, allowing the relay RL common movable contact C to move from NO to NC, connecting it to the low L tap on

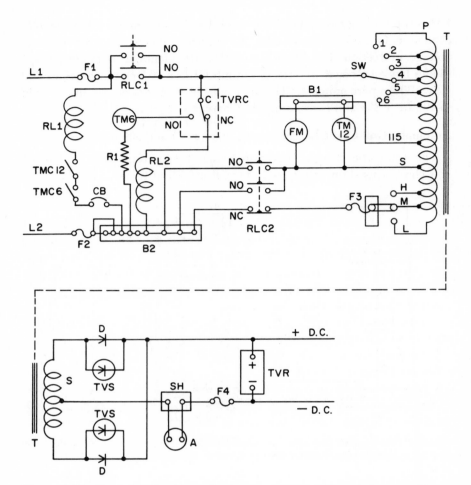

Figure 1–35 Typical 230-volt, single-phase automatic industrial battery charger, source 19 Model 1106 "C" case

the transformer T. The unit continues to operate on the low charge until the timer motor TM opens the timer motor contacts TMC. On some series of these models, the timer motor TM is connected to the NO contacts of the TVRC as shown by dotted line Y, preventing the timer from running until after the TVR cuts off at the full charge voltage. On other series of these models, the timer motor starts running from the time the charger is turned on, through dotted line X. Either line X or Y is used, but not both.

The bus-bars B1 and B2 are terminal strips with quick-connect terminals. Notice that the positive D.C. output lead is white and the negative output lead is black, which is opposite to the common wiring code of having the hot wire black and the ground wire white. The terminal voltage relay TVR and the terminal voltage relay contacts TVRC, are actually mounted on the same board. See Chapter 17, Section III, for details.

Automatic Industrial Chargers Models 1106 and 1112 "C" Case Single Phase, Model "D" Case Three-Phase

Figure 1-35 shows a 230 volt single-phase unit using one transformer. Figure 1-36 shows a 230 volt single-phase unit using two transformers with the two outputs in parallel. Figure 1-37 shows a 230 volt three-phase unit using three transformers. These are automatic heavy duty industrial type chargers that all operate on the same general principle, using the two-rate charging system.

Initial start-up procedure:

1. Turn "total hours of charge" timer to "on" (timer on the right side, when facing charger, TM12).

2. Turn left timer switch, TM6 to "normal". Fan will start and ammeter will show charging current.

3. Charger will run at high rate until gassing point of battery is reached (2.37 volts per cell at 77 deg. F); then, charge control (TVR relay) operates. If charger initially operates in low rate, push re-set button.

4. If low rate charging current is not about equal to the charging current specified by the battery manufacturer, proceed as follows:

   a. Turn timer to "off".
   b. Disconnect batteries and open disconnect switch.
   c. Open service door and move link on terminal board to high H, medium M or low L, as required.

Timer TM6 is a 6 hour timer, and timer TM12 is a 12 hour timer. Timer TM12 runs continuously as soon as the charger is turned on, and timer TM6 starts to run when the TVR operates to the NO contacts. Since contacts TMC12 and TMC6 are in series with the main line contactor,

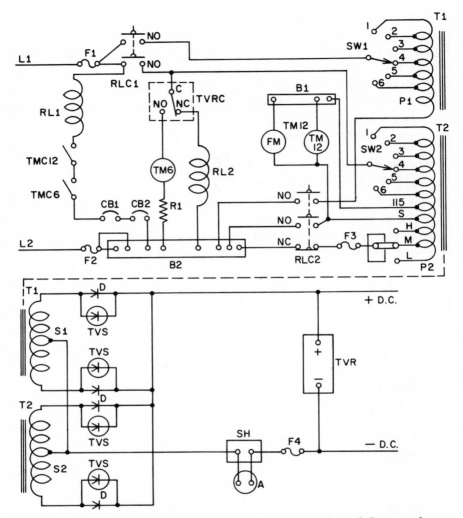

Figure 1–36 Typical 230-volt, single-phase automatic industrial battery charger with two transformers, source 19 Model 1112 "C" case

either one will shut the unit off. The circuit breakers CB are internal to each transformer, are connected in series, and each opens the circuit through the coil RL1, opening contacts RLC1 and disconnecting the A.C. input. For details on the TVR, see Chapter 17, Section III.

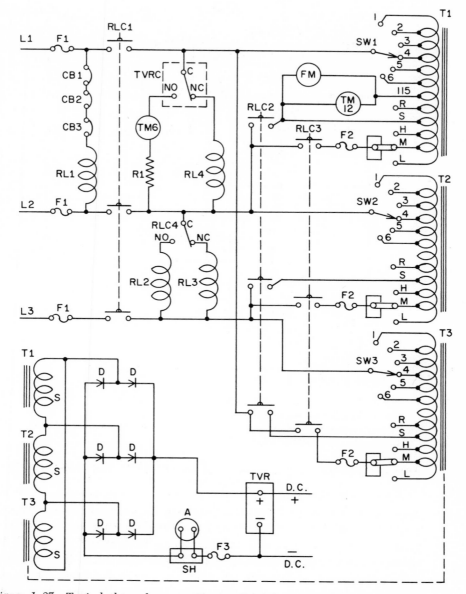

Figure 1–37 Typical three-phase automatic industrial battery charger, source 19
Model "D" case

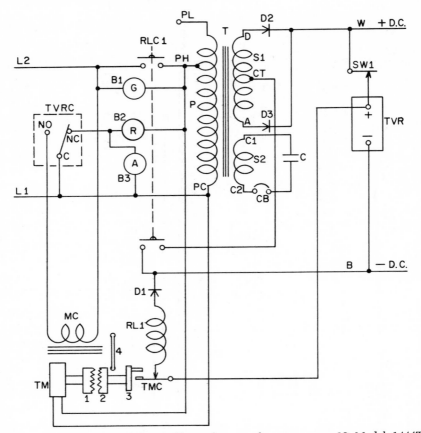

Figure 1–38 Typical automatic golf-cart battery charger, source 19 Model 1444T
(24V) and Model 1555T (36V)

Automatic Golf-Cart Charger Model 1444T (24V) and 1555T (36V)

Figure 1-38 shows an automatic golf-cart battery charger. With the
A.C. line and batteries connected, the charger operates at full rate charge,
and the red pilot light (high rate) B2 glows. When the battery becomes
fully charged, the terminal voltage relay TVR switches the TVRC,
opening the NC contacts and closing the NO contacts. The red pilot
light B2 goes out, and the amber (low rate) pilot B3 lights. These two
lights (red and amber) are now in series, but only the amber light glows
because it is a neon light, whereas the red light is of the heavier current

filament type. At the same time, the circuit is completed to activate the magnetic clutch coil MC, which starts a "low rate" timing cycle. The timer motor TM runs continuously driving toothed piece 1. When coil MC is energized, it attracts armature piece 4, causing piece 2 to engage with piece 1, thus turning piece 3 which is spring-loaded. At the end of the cycle (about 5 hours), the protruding stud on rotating piece 3 engages TMC arm and opens timer contact TMC. The contacts RLC1 also open the A.C. and D.C. circuits, and the green (charge complete) pilot light B1 glows, since it is connected across the A.C. set of contacts RLC1. The red and amber lights are out.

To test the operation of the finish cycle without waiting 5 hours, pull the A.C. plug, which releases the magnetic clutch and turn piece 3 with a finger until it is almost ready to operate switch TMC. While holding piece 3 in this position, plug in the A.C. plug, which operates the magnetic clutch and holds it in that position. It should open switch TMC in a few minutes.

The "low rate" cycle is misnamed, at least on the late models, because it does not reduce the charging current, except the normal taper rate as the battery approaches full charge. It is possible that the early models actually reduced the charging current on the "low rate" cycle by using the primary tap PL which is a low rate tap. The latest series do not have the PL tap.

The charger will turn on automatically when the battery voltage reaches the "storage on" setting of the TVR.

Pressing the NC momentary contact equalizer switch SW1 opens the D.C. circuit to the TVR, restoring the TVR contacts TVRC to the NC position for charging at the high rate to equalize the battery. Also, the magnetic clutch MC is released, and the low rate timer spring-loaded arm piece 3 returns to its starting position.

The small diode D1 located on the contactor RL1 prevents the contactor RL1 from closing on reverse polarity.

The transformer T is of the resonant type having a condenser C across a separate secondary S2. It is not a constant-current type, but it does have good voltage regulation for varying line voltages.

There is a thermal overload, circuit breaker CB, built into the transformer on the later models. It is in series with the secondary S2 and condenser C. It disconnects the condenser if the transformer overheats, which causes the D.C. output to drop considerably. When the transformer cools, the condenser is automatically reconnected and the charging resumes at full rate.

A lower than normal output usually indicates an open condenser or secondary S2, if the diodes D2 and D3 are up to standard.

See Chapter 17, Section III, for TVR details, and Chapter 11, Section III, for further details of timer construction.

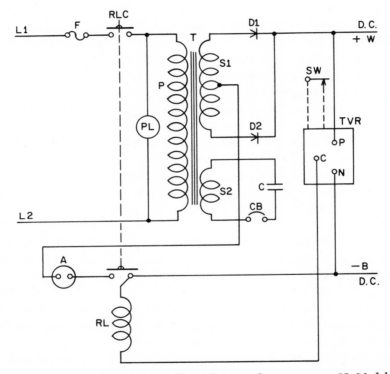

Figure 1–39 Typical automatic golf-cart battery charger, source 19 Models 1444 (24V) and 1555 (36V)

Automatic Golf-Cart Charger Model 1444 (24V) and 1555 (36V)

Figure 1-39 shows an automatic golf-cart and personnel carrier battery charger similar to that in Figure 1-38, except that it does not have a timer, and only one pilot light. The diode D1 shown in Figure 1-38 is mounted in the TVR of Figure 1-39. Switch SW shown in dotted lines is an equalizer switch used on some series of these models. See Chapter 17, Section III, for TVR details.

Multiple Battery Charger Model 36-100 (BC306) and 36-401

The charger shown in Figure 1-40 is a multiple battery charger having a "coarse" tap switch SW2 in the secondary circuit, and a "fine" adjustment in the form of a 0.5 ohm, 10 amp., 50 watt wire wound rheostat R.

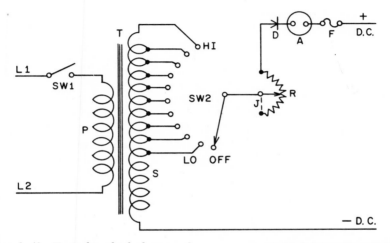

Figure 1–40   Typical multiple battery charger, source 17 Model 36-100 (BC-306) and Model 36-401

Usually, there is a third terminal on the rheostat as in a potentiometer. Always connect a jumper wire J, as shown in dotted lines, between the middle terminal and the un-used terminal to increase the life of this rheostat. This jumper wire J prevents arcing, and carries the current in case the wiper is lifted from the wire element by dirt or other foreign matter. D is a silicon diode; SW1 is an off-on switch.

Automatic Golf-Cart Charger Model VT24 (24V) and VT36 (36V)

Figure 1-41 shows an automatic golf-cart battery charger using two timers and a voltage regulator.

Transformer T1 is the main charging transformer, and is provided

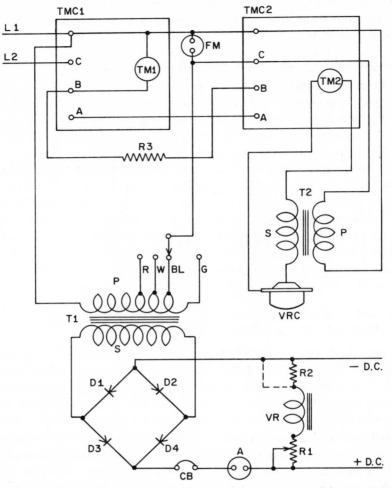

Figure 1–41 Typical automatic golf-cart charger, source 4 Models VT24 (24V) and VT36 (36V)

with four voltage taps, color coded as follows: green—125 volts, blue—120 volts, white—115 volts, and red—110 volts. If the output is too high (more than 25-30 amperes after ½ hour of charge), move the lead to a higher voltage tap. Transformer T2 is an isolation transformer.

The voltage regulator coil VR actuates the voltage regulator contacts

VRC. The voltage regulator is similar to automobile voltage regulators, and is temperature compensated. The same voltage regulator is used on 24 volt and 36 volt chargers, but the 36 volt charger has an additional voltage dropping coil R2. The dotted line around R2 shows the shorting wire connection for 24 volts. Potentiometer R1 provides voltage regulator adjustment. The operation of the charger is as follows:

1. With both controls in "off" position, connect A.C. cord to 115 volt line.

2. Connect D.C. polarized plug to cart receptacle.

3. Set voltage control TM2 to "start" position. Set time control TM1 to the "11" position. Both controls must be set for the unit to operate.

Setting the timer TM1 to "11" gives 11 hours of fast charging, followed by 6 hours of slow charge through slow charge resistor R3, unless the voltage regulator operates at an earlier time. The charger indexes from fast to slow charge about 30 minutes after batteries reach the voltage control point, and indexes to "off" after slow charge remains at voltage control point for 15 minutes. To slow charge only, set timer TM1 to zero for a slow "topping" charge. For timer details see Chapter 11, Section III, and for voltage regulator details see Chapter 17, Section III.

Golf-Cart Charger, Model 24S2 (24V) and 36S2 (36V), Also Model 24ST12 (24V) and 36ST12 (36V)

Figure 1-42 shows a golf-cart battery charger circuit which is rather unique, in that it uses a saturable reactor SR in the D.C. output positive lead. The saturable reactor allows full initial charging current of 25 amperes when the reactor is operating in the saturated region of its magnetic curve. As the charging current drops below this value, due to normal taper charge, to about 15 amperes, then, the reactor comes out of the saturation region and becomes reactive, offering a higher impedance to the pulsating D.C. current. This causes the charging current to fall off much more rapidly than would occur during the normal taper charge. The charging current thus drops very rapidly to a finish charge of less than 5 amperes. The reactor, consisting of an iron core and a coil, normally offers a high impedance to A.C. or to a pulsating rectified D.C. current. However, if there is a D.C. current between 15 and 25 amperes or more flowing through the winding the iron core will be saturated with flux, which cannot be increased very much at currents above this point. Therefore, it is almost like having no iron core at all in the saturated region. The impedance of the coil is almost the same as the low D.C. resistance of the coil. However, at currents below 15 amperes, the iron core is unsaturated causing the coil to offer a higher impedance. The impedance becomes even higher as the current is reduced, so the charging current drops very fast to a low "finish" value. TMS is a mechanical time switch. The charger is adjusted for various average line voltages by moving a lead to the proper tap.

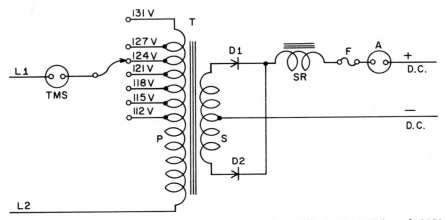

Figure 1–42 Typical golf-cart charger; source 15 Models 24S2 (24V) and 36S2 (36V) and source 16 Models 24ST12 (24V) and 36ST12 (36V)

Some chargers, especially the larger industrial chargers, may use a transistorized voltage control that feeds D.C. current through a second coil on the saturable reactor to control the degree of saturation and the charging current as determined by the battery voltage or state of charge, giving smooth control, and maintaining just the right charging current to hold the battery at full charge without the use of relays or contactors.

Constant-Current Multiple Battery Charger Model 6068

Automatic slow chargers, or multiple battery chargers, that will charge any voltage battery, or combination of batteries in series, without any switching, except possibly a "high" and a "low" rate switch, use a constant-current, leakage-reactance transformer and resonant secondary circuit to maintain a 6 ampere charging current, regardless of the number of batteries in series.

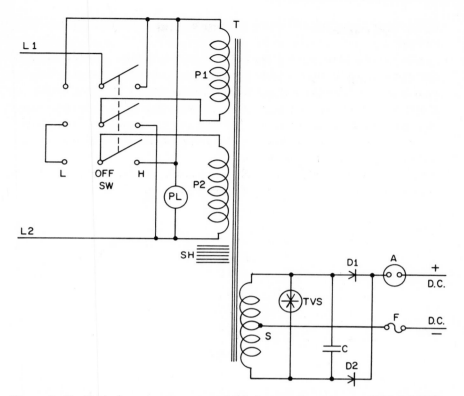

Figure 1–43   Typical constant-current, multiple battery charger, source 7 Model 6068

Figure 1-43 shows a multiple battery charger, which uses a leakage-reactance transformer T having a secondary S that is in resonance with condenser C. It does not have a separate resonant circuit secondary, and SH is the magnetic shunt.

There are two equal primaries P1 and P2 connected to the 115 volt A.C. line L1 and L2 through a 3PDT switch SW to provide "high" H and "low" L charging positions, and an "off" position. In the H position, the two primaries P1 and P2 are connected in parallel across the line. In the L position, the two primaries P1 and P2 are connected in series across the line. Pilot light PL is connected across the line in both H and L positions of switch SW. In the "off" position of switch SW, the primary windings and pilot light are disconnected.

A transient voltage suppressor TVS is connected across the secondary

S to protect the diodes D1 and D2 against high voltage "spikes" during switching.

The circuit, not only automatically compensates for a wide range of line voltage fluctuations without a change of charging current, but also maintains a constant charging current throughout the charging cycle. The charging current does not taper off to a lower rate, like a constant voltage charger does. More charging can be done in a given period of time, which can be calculated by multiplying the amperes times the hours to determine the ampere-hours of charge.

Small capacity batteries, such as motorcycle batteries, should be charged with the charging control switch SW in the "low" L position.

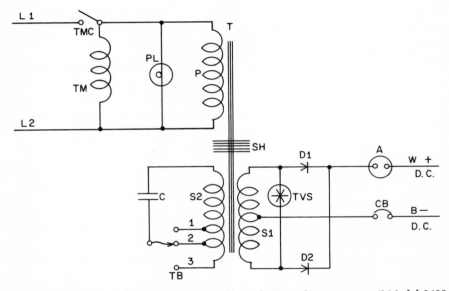

Figure 1-44 Typical constant-current golf-cart battery charger, source 7 Model 2400 (24V) and Model 3600 (36V)

Constant-Current Golf Cart Charger Model 2400 (24V) and 3600 (36V)

The circuit shown in Figure 1-44 is for a 24 volt or 36 volt golf-cart charger using a transformer T having a resonant secondary S2 and condenser C. Three adjustments 1, 2, and 3 on terminal board TB are provided so that a replacement condenser, if not exactly of the same mfd as the original, can be matched to the transformer. Use the tap giving the highest voltage across the condenser, which should be about 400 volts.

The output is constant, and the timer TM can be set for 3 hours of charge for each hour of machine use. Once every week, whether or not the machine is used, set the timer TM to the 12 hour mark and complete a full charge.

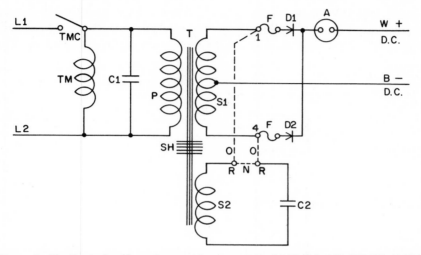

Figure 1–45 Typical golf-cart battery charger, source 16 Models 18LC25T12 (18V), 24LC25T12 (24V), and 36LC25T12 (36V)

Golf-Cart Charger, Models 36LC25T12 (36V), 24LC25T12 (24V) and 18LC25T12 (18V)

Figure 1-45 shows the circuit of a resonant transformer T with secondary winding S2 and condenser C2. On early models, the two red R leads from one side of the condenser C2 and one side of transformer secondary S2 were connected as shown by dotted lines 0 and 0 to terminals 1 and 4, placing S2 and C2 in series with each other and in parallel with the secondary S1. On later models, the two red R leads are connected together as shown by dotted line N, placing S2 and C2 in series and isolating them from the rest of the circuit. Also, on later models, a condenser C1 (0.5 mfd-600V) was added for voltage surge protection.

These models have a taper charge, so they are not of the constant-current type. But, the resonant circuit does provide full output over a wide range of line voltage fluctuations.

Bulb-Type, Half-Wave Multiple Battery Charger

Multiple battery chargers charge several batteries connected in series at an overnight, or 8 hour, rate of 6-12 amperes. They will usually charge 1-12 6-volt batteries or 1-6 12-volt batteries, or any combination of 6- and 12-volt batteries up to a maximum of 72 volts. Some chargers charge up to only 36 volts, while others may charge over 72 volts. A typical half-wave, bulb-type multiple battery charger is shown in Figure 1-46. The

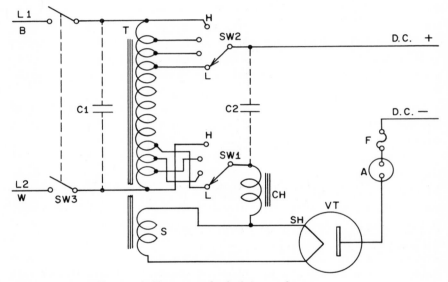

Figure 1–46   Bulb-type, half-wave multiple battery charger

auto-transformer T shown requires either a permanent connection to the power line, with the white W wire going to ground, or a polarized plug. Even then, there is some danger if a battery clip should touch a grounded object. The auto-transformer often has a DPST switch SW3 to break both sides of the A.C. line. These chargers are also available with a separate primary winding, using a cord and plug. Also, a SPST switch is sufficient as an off and on switch.

Although the tungar type bulb VT is shown, a silicon diode is used often in this same circuit. When a silicon diode is used, the choke CH and the filament secondary S are not needed. Some silicon diode circuits use a resistor in the same position as choke CH. Sometimes, either or both condensers C1 and C2, shown in dotted lines, are added as transient voltage suppressors, and are usually about 0.5 mfd, 300V.

Rotary switch SW1 is used for the "coarse" adjustment, and rotary switch SW2 is used for the "fine" adjustment of battery charging current. Usually, the coarse switch SW1 is marked in the number of batteries connected in series, such as 1-2, 3-4, and the fine switch SW2 is adjusted to give the desired rate of charge as shown on the ammeter A. If switch SW1 is not marked, always start in the lowest position (CCW) and run switch SW2 through all positions until a reading is obtained on the ammeter. If no reading is obtained, turn switch SW2 back to the low position (CCW). Then, turn switch SW1 to the next higher position (CW). Adjust the two switches SW1 and SW2 until the desired current is obtained. Only four taps are shown for each switch, but there are usually more.

The connection from the tungar bulb to the choke CH should be on the shell SH side of the tube socket, as it must carry both the charging current plus the filament current.

A resistor of about 1 ohm, or a choke, must be used in the position of the choke CH to stabilize the operation of the gas-filled tungar bulb. Sometimes a choke CH is a separate coil of wire wound on a separate iron core, or it may be wound on the transformer core as a separate winding. If the choke is burned out, it must be replaced, rewound, or replaced with a resistor. A simple solution is to remove the choke or the wire around it, and replace the tungar bulb with a screw-in type of silicon diode replacement unit, which does not require a choke. However, a resistor may have to be added in the place of choke CH, if the current on the lowest battery voltage cannot be reduced.

Twin-Bulb, Full-Wave Multiple Battery Charger

A full-wave tungar bulb type of multiple battery charger, using two bulbs VT1 and VT2 and an auto-transformer T, is shown in Figure 1-47. However, it may be built with a separate insulated primary for 115 volts and one for 230 volts. The 230-volt chargers always have a separate primary winding.

Switches SW3 and SW4 are the "coarse" switches, and are ganged

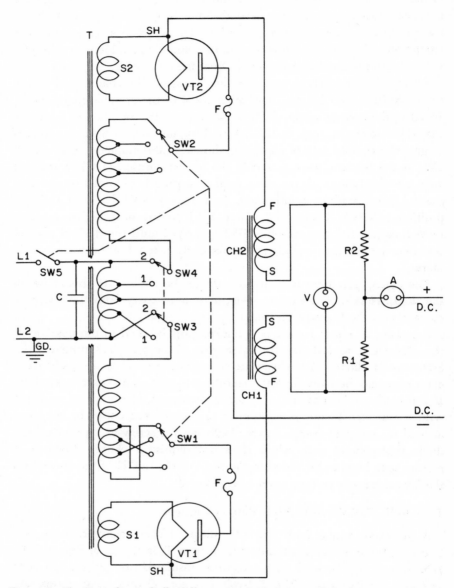

Figure 1–47 Typical twin-bulb, full-wave multiple battery charger

together. In position 1, the charger will handle 1 to 6, 6-volt batteries connected in series, and in position 2, it will handle 7 to 12, 6-volt batteries connected in series. Switches SW1, SW2, and SW5 are ganged together. SW1 and SW2 are the "fine" adjustment, and SW5 is the line switch to disconnect the A.C. when in the "off" position. Only 4 taps are shown for SW1 and SW2, but usually there are about 11 taps.

Most twin-bulb, full-wave chargers have a milli-voltmeter V in addition to the ammeter A to show if both bulbs are operating. It is usually called a "bulb indicator" and operates in the middle of the scale at zero voltage when both bulbs are operating normally. If one bulb is inoperative, the meter V will read to the right or left, showing which bulb is not operating. The causes for one bulb failing to operate could be a loose anode connection, or other loose connection, loose bulb or burned out filament. The meter V is operated, to the right or left, by an unbalance of the currents through chokes CH1 and CH2 and resistors R1 and R2. Chokes CH1 and CH2 are wound on the same iron core, and are needed to stabilize the operation of the bulbs. With both bulbs operating, the current from bulb VT1 goes through choke CH1 and resistor R1 and out the positive D.C. to the batteries. The current from bulb VT2 goes through choke CH2, resistor R2, and out the positive D.C. to the batteries. If the currents through these two circuits are exactly equal, there is zero voltage across meter V, and it will read zero in the middle of the scale. If there is current from only one bulb, there will be a voltage drop across either R1 or R2, causing the meter V to move either to the right or left of middle scale. It is important to have the correct winding connections to chokes CH1 and CH2, as indicated by start S and finish F of the winding. A loose connection on the terminal block that carries the choke, resistor and meter leads, can cause an unbalanced reading. R1 and R2 are very low resistance shunts.

If the chokes CH1 and CH2 burn out, they must be replaced or rewound if bulbs are used. If a screw-in type of silicon diode replacement is used instead of the bulbs, the chokes can be eliminated, or a 1 ohm resistor substituted.

Condenser C is a 2 mfd, 300V unit for transient voltage protection.

The shell SH of each bulb socket must always be connected to the charger positive output because it carries the charging current plus the filament current. The shell SH has a heavier current carrying capacity than the center contact.

Battery Charger Circuit Combinations

All battery chargers consist of a primary circuit and a secondary circuit. Almost any combination of primary and secondary circuits will be found in modern battery chargers.

Some battery chargers have various automatic and semi-automatic controls in the primary circuit to disconnect the A.C. current, or reduce its value to provide a trickle charge indefinitely, or for a predetermined time, by means of a timer or a relay in the A.C. circuit actuated by conditions in the secondary or battery circuit. This secondary control is controlled by the battery voltage or condition of charge, or by a thermostat bulb placed in the battery cell electrolyte to cut off the charger at a predetermined temperature, usually about 125 deg. F. Other secondary controls may hold the charger inoperative if the charger clamps are reversed on the battery post, commonly called an alternator protector or reverse polarity protector.

Also, there are primary and/or secondary circuit breakers that disconnect the charger if the maximum safe current is exceeded. Some are automatically reset, and others are manually reset. Fuses may also be found in the primary and/or the secondary circuit for the same reason as circuit breakers.

Some chargers are completely automatic, maintaining a charge on a battery, starting and stopping as the battery condition changes. These devices use many components that are explained in detail in subsequent chapters.

Dry Cell Battery Charger

Although zinc-carbon dry cells are not considered rechargeable, they can be rejuvenated successfully many times if charged while new or only partially discharged. Figure 1-48 shows a typical charging circuit for this purpose. Switch SW1 is a DPST safety switch that opens when the lid is lifted to prevent shock hazard. The charger operates only when the lid of the case is closed. Current limiting and charge indicating bulb B1 is a 10 watt, 115 volt incandescent lamp #10C7, or equivalent; diode D1 is a 750 ma, 400 PIV silicon diode; switches SW2, SW3, SW4 and SW5 are normally closed to complete the series circuit, but open when batteries are inserted, so that the current goes through the batteries.

On one model, there are three compartments, one for two size AA or C cells, one for two size D cells, and one for a 9 volt radio battery. But these chargers are made for any number or combination of cells.

As in any transformerless half-wave rectifier circuit, this charger can be used on 115 volts of any frequency, such as 60Hz, 50Hz, or 25Hz, or even on D.C. On D.C., the polarity is correct when the bulb B1 lights because the diode D1 will conduct in only one direction.

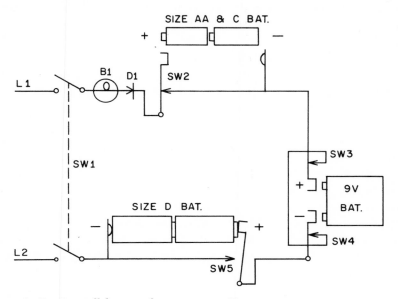

Figure 1–48 Dry cell battery charger, source 11

This circuit charges at a rate of approximately 45 ma, and uses about 5 watts regardless of the number of cells.

The approximate charging time is 4-6 hours for the AA penlite cells; 10-16 hours for size C; 12-18 hours for size D cells; and 5-7 hours for 9 volt radio batteries. Twice this time should be allowed for nickel-cadmium cells. Silver and mercury batteries should not be charged on a circuit using these component values.

The charge indicator bulb B1 should light at all times with the lid closed, but it does not necessarily indicate that the batteries are being charged. It only shows that the series circuit is complete. For batteries that may be shorter than standard, the contacts may not open, and would not receive a charge even though the light is burning. These contacts must be open when the cells are inserted. A piece of aluminum foil, folded or wadded, can be placed between the base of one cell and the holder to push the cell toward the contact switch to open it.

The size D cells are shown installed for charging, with switch SW5 open. The size AA & C are shown not installed for charging and switch SW2 is closed. The 9V radio battery is shown not installed for charging, and switches SW3 and SW4 are closed.

SECTION II

TEST EQUIPMENT REQUIRED 1

It does not require a lot of time to test and repair battery chargers, if the proper equipment and tools are available and the test procedures are understood.

Current Limiting Input Tester

Before testing any battery charger, plug it into a current limiting outlet to prevent blowing fuses in case of a short circuit. Such a current limiting input test set can be made by following the schematic wiring diagram shown in Figure 1-1.

CB1 is a 20 ampere manual reset circuit breaker of the fast acting type; F1 is a 20 ampere fuse; SW2 and SW4 are SPST toggle switches rated at 15 amperes at 115 volts each; the bulbs R2 and R3 are 250 watt, 115 volt infrared heat lamps (a single 550-600 watt cone type heating element can be used); R1 is a 100 watt, 115 volt light bulb; V is a 0-150 V.C.A. voltmeter; A is a 0-15 ampere ammeter; SW3 is an ammeter-shunting switch rated at 15 amperes at 115 volts, with normally closed contacts that must be held open to read the ammeter A. The outlet receptacle REC is a "U" grounded outlet for plugging in a charger A.C. cord, or for plugging in the test cord TC. The test cord TC is a 2-line cord with two small insulated alligator clips for testing transformers, timers and motors separately.

Switch SW2 is shown in the "in" position, which means that the high-resistance lamp R1 only, or R1, R2 and R3 are in the circuit in parallel. The "out" position of switch SW2 means all resistance is shorted out of the circuit and full voltage is applied directly to the receptacle REC.

The switch SW4 is shown in the "H" position indicating that the high resistance lamp R1 is in the circuit only if switch SW2 is in the "in" position. The switch SW4 in the "L" position means that low resistance, consisting of lamps R1, R2, and R3 in parallel, is in the circuit, if switch SW2 is in the "in" position.

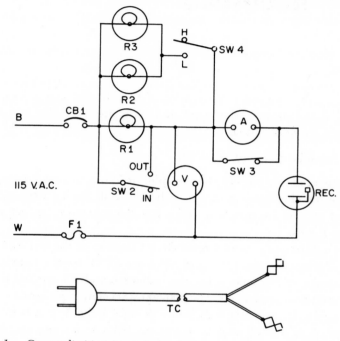

Figure 1–1   Current-limiting input test set

Before plugging the charger in the receptacle, set switch SW2 to the "in" position, and set switch SW4 either to the "H" position for testing small chargers, timers and motors or to the "L" position for testing larger chargers.

Plug in the charger but do not connect the D.C. cables to the battery, or allow them to touch each other. Turn the charger on, run through all of the rate and battery selector positions and watch the voltmeter V. The voltage will drop to zero or near zero if there is a short circuit anywhere in the primary or secondary circuit. If the voltage doesn't drop at all, there is an open circuit in the primary circuit or line cord. If there are no shorts or opens, the voltmeter will read anywhere below the line voltage to as low as 30 volts for the larger chargers that have a normal primary current of 10 to 15 amperes.

If there are no shorts, set switch SW2 to the "out" position, connect the battery cables to the battery, and proceed to test the charger. To read the amperes input to the charger, press switch SW3 and read the ammeter A.

For further test procedures on battery chargers, refer to Chapter 2, section IV.

The circuit of Figure 1-1 can be assembled and mounted on a "bread board" panel or built into a well-ventilated metal cabinet. All of the parts are readily available locally, from electrical supply houses.

This input tester can be connected to the A.C. line continuously to monitor the line voltage if desired.

Volt-Ohmmeter

A volt-ohmmeter is an essential piece of equipment, but need not be as elaborate or as sensitive as required for radio or TV service. However, it should be as accurate as possible. The important ranges should include:

$$
\begin{array}{ll}
\text{D.C. Volts} & \text{2.5-10-50-250} \\
\text{A.C. Volts} & \text{2.5-10-50-250-1000} \\
\text{Ohms} & \text{1-10-100-10K-100K}
\end{array}
$$

A volt-ohmmeter that meets these requirements is the Simpson 260 Series 5, available from many electrical supply stores.

Capacitor Analyzer

A capacitor or condenser analyzer is valuable for testing a large variety of capacitors and condensers used in battery charger circuits. It should be able to test oil and paper condensers and electrolytic capacitors for actual mfd value, leakage, and power factor. The important ranges should include oil and paper condensers 0.001-100 mfd and electrolytic capacitors 1-1000 mfd. These analyzers are readily available in kit or wired form from various electrical supply stores.

Rectifier and Diode Tester

A rectifier and diode tester is a big time saver. There is no known tester available that will test both heavy current diodes and rectifiers as well as small low current diodes, or that will indicate polarity of rectifiers, diodes and transistors. For heavy current rectifiers, a heavy-duty unit is available at electrical supply stores or can be purchased from auto supply houses as alternator diode testers.

A good rectifier and diode tester can be made from readily available materials by following the details of Figure 1-2. This tester will test all rectifiers and diodes, as well as power transistors, for opens, shorts, conductivity, and polarity, and will test the polarity and condition of other low current solid state devices rated at 40 m.a. or over. The main rectifiers in battery chargers can, in most cases, be tested for conductivity, shorts and opens while connected in the circuit without disconnecting the transformer. Other small diodes and transistors may have to be disconnected for testing.

In Figure 1-2, the transformer T1 can be any battery charger transformer with a tapped secondary able to charge at a 6 ampere rate on 6 or 12 volts. The primary 6 volt tap and the common tap are used as the two outside leads. Switch SW1 is a SPST on-off toggle switch rated at 3-6 ampere at 115 volts; switch SW2 is a 3 contact rotary switch rated at 6 amperes at 115 volts for calibrating the two meter ranges: position H to calibrate rectifiers and diodes rated over 5 amperes, and position L to calibrate rectifiers and diodes rated under 5 amperes. Position T is for testing conductivity through prods PR3 and PR4; switch SW3 is a DPDT switch with no off position rated at 6 amperes at 115 volts that is used to select the high and low current ranges: L position is for rectifers rated under 5 amperes drawing about 100 m.a. current, and position H is for rectifiers rated over 5 amperes and drawing about 5 amperes. Potentiometer R1 is a 100 ohm, 2 watt wire wound resistor with a linear taper at the low range under 5 amperes. Potentiometer R2 is a 2 ohm, 50 watt wire wound resistor with a linear taper for calibrating the high range above 5 amperes. Potentiometer R3 is a 25 ohm, 1 watt wire wound linear taper resistor for calibrating both ranges. Resistor R4 is a 20 ohm, 1 watt fixed resistor.

Meter V is an expanded range 6 volt voltmeter. Any 6 volt battery

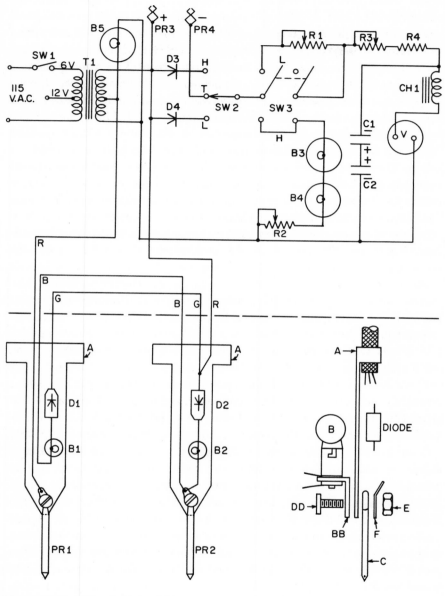

Figure 1–2   Rectifier and diode tester

tester voltmeter can be used if resistors R1, R3, and R4 are correct for the substitute meter. An expanded range voltmeter gives the accurate readings needed to find the rectifiers with only 80% conductivity since they are useless in battery chargers. Capacitors C1 and C2 are each 100 mfd, 25 volt electrolytic capacitors connected back to back as a non-polarized unit, if a rectifier under test is connected for reverse polarity. Choke CH1 is a 7 ohm primary transformer winding taken from a 6 volt, 4 ampere charger. Choke CH1 and capacitors C1 and C2 filter the rectifier output, holding the meter needle steady. Bulbs B3 and B4 are number 2330 6-8 volt auto headlight bulbs with their filaments connected in series. These bulbs have a flange which can be used for mounting and connecting the two bulbs together, while the filaments in each bulb are connected together by soldering the lead across both terminals. Diode D4 is a silicon diode used for loads under 5 amperes and can be any small diode with a rating of 750 m.a., 200 PIV. Diode D3 is a diode for loads over 5 amperes and can be any 35 ampere, 50 PIV silicon diode, such as Motorola IN1183. Pilot light B5 is a No. 53 bulb, or any 12-14 volt bulb.

The leads to each of the test prods PR3 and PR4 should be about 4 or 5 feet long, or long enough to reach from the test bench to the battery charger under test, and may terminate in alligator clips or pointed prods. Test prods PR3 and PR4 are used for conductivity and quality tests on all rectifiers and diodes used in battery charger circuits. Although they will indicate polarity, shorts and opens, prods PR1 and PR2 are better for determining all these conditions except quality and are used first in testing rectifiers and diodes. Prods PR1 and PR2 are mechanically the same but are wired differently as shown in Figure 1-2. Diodes D1 and D2 are both the same as D4. Bulbs B1 and B2 are both No. 53 type lamps. Sockets BB are alike and are single contact miniature bayonet with offset bracket insulated from both terminals. A copper sheet cut as shown as part A has two tabs on one end that fold over as a cable clamp, and a hole drilled in the opposite end. Part C is a No. 12 copper wire ground to a point on one end and an eyeloop formed on the other end for mounting. Parts A, BB, C, and solder terminal F are held together by a 6-32 X ⅜ inch machine screw DD and a 6-32 nut E. After complete assembly and test, wrap the whole prod with plastic electrical insulating tape except the glass part of the bulb for viewing. The three-wire, extra flexible cable connecting the two prods PR1 and PR2 to each other and to the tester shows a black B, green G, and a red R color coded cable.

When using the rectifier tester, prods PRI and PR2 are used to determine the polarity and to indicate shorts or opens in rectifiers, diodes and transistors. These prods will show whether or not the diodes or rectifiers are working but not how well they conduct. When prods PR1 and PR2 are placed across any rectifier, diode or transistor section capable of carrying 40 m.a. or more, regardless of how the prods are connected to the terminals of the unit being tested, only the bulb B1 or B2 associated with the prod on the anode of the unit will glow indicating rectifier action. If neither bulb glows the unit is an open circuit. If both bulbs glow, the unit is shorted. If both bulbs glow unevenly there is some shunt leakage, such as occurs during "in circuit" testing, or the rectifier under test has almost as low a backward resistance as forward resistance which is reason enough to reject the unit. If the unit being tested is shorted or open, do not continue to test for conductivity with prods PR3 and PR4.

Prods PR3 and PR4 test rectifiers, diodes and power transistors capable of carrying 100 m.a. or over for conductivity and quality, also confirm shorts or opens.

Initially, and periodically afterward, it is necessary to calibrate meter V to the built-in calibration standards of diodes D3 and D4. To calibrate, plug in line cord of tester and close switch SW1; set SW2 and SW3 to position L; set potentiometer R3 in mid range, and adjust potentiometer R1 until meter V reads at the far right edge in the green "SAFE" scale under "CHARGE." Ignore the two upper scales "LOAD TEST" and "GEN REGUL." If unable to reach this point, adjust both R3 and R1 until it is reached. Bulbs B3 and B4 do not glow on this setting. Next, turn switch SW2 and switch SW3 to H position (bulbs B3 and B4 should glow) and adjust R2 until the meter V reads at the far right edge of the green "SAFE" area. It may be necessary to repeat these two adjustments, readjusting R1, R2, and R3 accordingly, until the meter reads the same on both ranges. No diode or rectifier tested should exceed this point if diodes D3 and D4 are up to standard. All good rectifiers, even the older low conductivity selenium rectifiers, should fall well within the "SAFE" range. Border line cases can be decided by giving the charger an output load test.

After the meter is calibrated, the following test procedure should be done. First, use prods PR1 and PR2 to test for opens, shorts, and polarity. If there are no opens or shorts, the bulb in either prod PR1 or prod PR2 will glow, indicating the anode. The red test prod PR3 should be connected to the anode terminal, and the black prod PR4 should be connected to the cathode terminal. Set switch SW3 to either H for over 5 amperes rating, or to L for under 5 amperes rating, and set switch SW2 to test position T. Turn on switch SW1, and the meter should read in the green "SAFE" area if the unit is acceptable. On the H position of SW3, a shorted unit will cause both bulbs B3 and B4 to increase in brilliancy. Be sure the polarity is correct. A reversed connection would show no reading on the meter.

Do not leave the tester on or connected for calibration or testing any longer than is necessary to obtain a reading.

This tester can be used to test power transistors only on the low range L below 5 amperes which draws 100 m.a. current.

Prods PR3 and PR4 draw a current through the rectifier of 5 amperes on the H range and 100 m.a. on the L range. Prods PR1 and PR2 draw a 40 m.a. current through the rectifier.

Battery Charger Load Tester and D.C. Welder

Every battery charger output should be tested, under load, for the correct rated output at rated line voltage.

A good load tester, which can be purchased from electrical supply houses, can test all batteries from 2 to 40 volts and, when used with batteries, can discharge or hold the batteries at the desired voltage. It uses a carbon pile rheostat to adjust the load to any desired value, using its 3 range voltmeter and ammeter. Any battery load tester can be used but they are usually limited to 6 and 12 volts.

A load tester can be built to load test at various voltage levels using the schematic shown in Figure 1-3. The most useful load test voltages are 6, 12, 18, 24, 30 and 36. If only 6 and 12 volt units are tested, the other sections can be eliminated, or more added if higher voltages are required. The six 6 volt batteries shown should all be the same and can be any automotive 6 volt lead-acid battery which need not be heavy duty because they will last for years in this service if kept charged. Larger batteries are required for industrial chargers having an output of over

100 amperes. These batteries will usually be kept charged by the process of testing chargers. However, they should be equalized occasionally by charging them often enough to keep the electrolyte up to full charge in all cells. A simple charger can be made using a transformer and rectifier from a 36 volt charger, adding any refinements desired. Resistors R1 are identical 6 volt battery test load resistors used in commercial battery load testers. They can be purchased as a new repair part, salvaged from old junked testers, or they can be made using the construction details of Figures 1-4 and 1-5. Make all sections the same as section A-B, that is, 12 feet of 5 strands of 0.0625 inch baling wire wound on an asbestos board, winding on additional sections if used for welding as explained later. Solenoids RL1 are the same and are standard 6 volt automotive intermittent duty starting solenoids having a single small terminal between the large studs and connected to one end of the coil. The other end of the coil is connected internally to the "BAT" terminal. The contacts to the large studs are normally open, but close when the coil is energized. Switches SW1 are the same and are momentary contact N.O. SPST pushbutton switches. Switches SW2 are alike and are SPST rated at 1-3 amperes at 115 volts. Switches SW3 are identical SPDT switches with an off position. Switches SW2 and SW3 should always be turned off when not in use because the meters would put a small drain on the batteries. Meter A1 is a 0-10 ampere D.C. ammeter. Meter A2 is a 0-100 ampere D.C. ammeter. Meter VI is a 0-8 volt D.C. voltmeter for 6 volt chargers; meter V2 is a 0-15 volt D.C. voltmeter for 12 volt chargers; meter V3 is a 0-30 volt D.C. voltmeter and is switched for reading either 18 or 24 volts; meter V4 is a 0-50 volt D.C. voltmeter that can be switched for reading either 30 or 36 volts. Voltmeters V1, V2, V3 and V4 should be accurate. If they are not accurate, they can be recalibrated for a single voltage by setting the zero adjustment if accessible on the face of the meter. Otherwise, the voltmeter can be calibrated by using a rheostat in series with the meter to reduce the reading if it is high, or a rheostat in parallel with the meter internal multiplier resistance to increase the reading. If an accurate standard is unavailable, calibrate the meter using a good fully charged lead-acid battery at normal room temperature of about 77 deg. F that has lost its surface charge, such as 24 hours after fully charging using a value of 2.1 volts per cell. The test terminals from the batteries extend to a convenient place on the front of the bench

and are marked as shown. The negative terminals are painted black and marked "NEG" or are left plain. The two positive terminals are painted red or marked "POS." A rubber curtain, such as a piece of old inner tube, should be hung over the terminals to prevent accidental shorts, and removed when the load tester is used. Make the test terminals using a piece of copper tubing about 3 inches long that has been soldered or crimped on the ends of the cables and flattened with a hole drilled for securing. Then, bend them outward away from the bench to take the charger clamp.

To use the load tester, connect the battery charger positive or red clamp on the proper positive test terminal. For small chargers under 10 amperes, use terminal positive 10 A., and for chargers above 10 amperes, use terminal positive 100 A. The negative battery charger clamp is connected to the correct voltage negative test terminal, such as 6, 12, 18, 24, 30 or 36.

Set the correct voltmeter switch SW2 or SW3 to match the voltage of the charger being tested. Plug in the charger A.C. cord, turn it on to charge, and note the ammeter reading on both the charger and on the test panel. They should read about the same. However, on many battery chargers, the ammeters usually read on the high side. That is, they show more amperes than the charger is actually putting out, which is safer than if the reading is too low. If the charger is putting out its full rated output on the highest setting and the D.C. voltmeter reads nominal voltage or higher, and if the line voltage is normal, the charger has full output.

If the D.C. current is not up to its rated output, the battery voltage has probably risen above its nominal value. Press the proper switch SW1 for the voltage of the battery to put a load on the battery and charger. Hold in switch SW1 until the D.C. voltmeter reads exactly the nominal voltage, such as 6, 12, 18, 24, 30 or 36. Then, quickly read the D.C. ammeter on the tester and charger. It may be necessary to turn off the charger and discharge the battery for 20 seconds to bring the battery voltage down.

The procedure is the same when using a ready-built load tester. Connect the tester clamps and the charger clamps to the battery, and apply the correct number of discharge load resistors.

The load tester is used to reduce the battery voltage to its nominal value. A charger will produce rated amperes only on a fully discharged

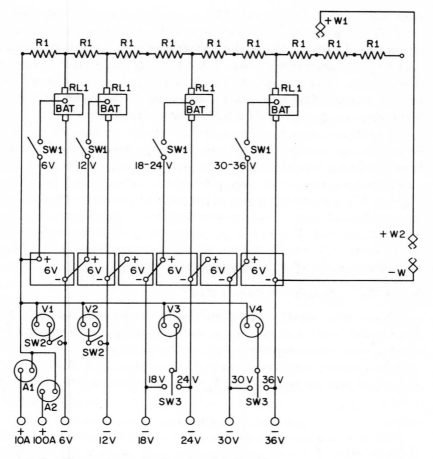

Figure 1–3   Battery charger load tester and DC welder

battery having a voltage at or below its nominal rating. As the battery is charged on most constant voltage chargers, its voltage increases and the amperes decrease, so the load resistance discharges the battery and helps to hold it down during testing.

A single resistance load tester is not practical in a repair shop because of the large variety of chargers tested. However, once the volt-amperes across a given resistance have been established, it is an accurate indication of the full output of the charger. Charging batteries is different from applying a voltage across a resistance only. The charger must first overcome the back EMF or counter voltage of the battery before it can deliver any charging current. After the charger reaches the nominal dead battery voltage, the current drawn by the battery is determined by the internal resistance of the battery. This resistance consists of the battery plates, straps, electrolyte and is a very low value in fractions of an ohm. That is why the conductivity of the rectifiers, and the line voltage are quite critical for full output of the charger.

The load tester shown in Figure 1-3 also makes an ideal D.C. welder for metallic arc welding, straight and reverse polarity, arc torch brazing, heavy soldering and heating. The voltage dropping resistor consists of all of the load resistors R1 plus additional sections. Only 3 are shown. Clip pos. W1 is connected to one of the junctions between any two of the resistors R1, and the pos.W2 clip is the positive welding electrode holder, or carbon arc torch. Clip neg. W is the negative welding connection. This arrangement will handle the occasional welding, brazing, or heavy soldering encountered in battery charger service.

Heavy Duty Adjustable Resistance Board

A heavy duty adjustable resistance board can be used, not only as a more flexible load tester and battery discharge resistor at all voltages, but also for many other purposes in battery charger service.

Make a resistor board as shown in Figures 1-4 and 1-5. With the cable and clip K and the jumper wire J, any load resistance can be selected

JUNCTION	A	B	C	D	E	F	G	H
STRANDS	5	4	4	2	2	1	1	
LENGTH	12'	7'2"	17'8"	23'2"	22'5"	46'	23'	
MAX. AMPS.	100	80	80	40	40	20	20	
MAX. VOLTS	6	6	12	12	24	36	36	
RES. OHMS	.048	.036	.089	.232	.225	.92	.46	
RES. TOTAL	.048	.084	.173	.405	.630	1.55	2.01	

Figure 1–4 Heavy duty adjustable resistance board winding data

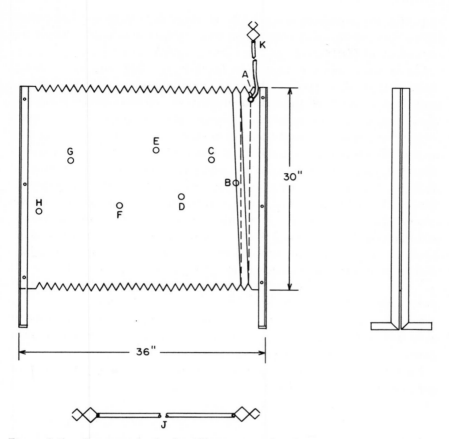

Figure 1–5· Construction of adjustable resistance board

to reduce or hold the battery voltage at any desired value. If desired, a heavy duty carbon pile rheostat can be connected in series to give a fine control, and a long coil of heavy nichrome heater wire can be added at junction H to give additional higher resistance. A solenoid, or foot operated automobile starter switch, can be used in series to connect or disconnect the load resistor quickly and conveniently.

The test board shown in Figure 1-5 is wound using iron baling wire having a diameter of 0.0625". The number of strands and length of each section are shown in Figure 1-4. These strands are twisted, after measuring and cutting to length, by fastening one end in a vise and the other ends to the spokes of a wheel. By spinning the wheel, the strands are twisted tightly.

The board is asbestos ¼ inch thick and 36 inches long by 30 inches wide, clamped on each end by bolts inserted through 1″ X 1″ X ⅛″ angle iron as shown. If this is to be used on an inflammable floor, such as wood, prepare a sheet metal pan and place it under the feet of the board to catch any sparks or molten metal in case of an accidental wrong connection. Cut notches, using a hack saw, in the edges on the top and bottom on 1 inch centers to keep the wires separated and in position. At each junction twist the wires together, bring them out at right angles to the board, and braze or solder the ends together to make good electrical contact. Junctions that occur on the back side of the board such as C and F, can be brought out to the front through a drilled hole in the board. Each junction should be secured by wire through small holes drilled in the board. Insulate any close wires using asbestos paper. Intermediate taps can be brought out after the board is completed by using two-piece copper connectors such as those used by electricians.

Figure 1-4 shows the maximum amperes, approximate resistance in ohms for each section, and also the total voltage and resistance from junction A to each successive junction.

To use the board, connect lead K to one side of the battery and charger, and connect jumper J between the other side of the battery and charger, and to the appropriate junction to give the desired load.

After some experience, the board can be calibrated for testing any charger without batteries. The voltage and/or the amperes output will be lower than when connected across a battery. Make up a chart for each model known to have full output on a battery by selecting a junction that will give approximately the full rated amperes and record the exact amperage, voltage and junction used, as well as the line voltage when the test board is cold at start. The D.C. voltage will be considerably below the nominal voltage. To test the same models, or a similar model having the same rating, using only the resistance board and no battery, the readings should be approximately the same. Remember that the resistance will rise and the current will drop as the resistance board heats up. A quick cold start test each time should produce the same readings.

Figure 1-4 shows the maximum voltage to apply between junction A and the other junctions. However, a lower voltage applied between junction A and a higher junction will draw a lower current. For example, 6 volts between junction A and F would draw approximately 10 amperes (0.630 ohms).

This board can be used for other purposes also such as the voltage dropping resistor shown in testing voltage regulators in Figure 1-6, resistor R1.

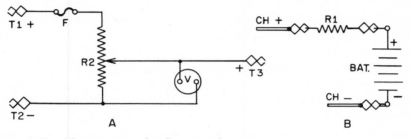

Figure 1–6   Charger-powered voltage regulator test set

The test resistance board can be used to determine the correct resistance of ballast or voltage dropping resistors used in primary or secondary circuits of battery chargers when connecting or converting old style 6 volt only chargers to 6/12 volt operation or to reduce the output in some cases. After determining which junctions give the proper value, check the resistance accurately on a low range ohmmeter or wheatstone bridge. For low values, calculate the resistance using Ohm's law after reading the amperes and the voltage drop across the resistor section. The resistance in ohms equals the voltage drop divided by the current in amperes. For example, with 10 amperes through the resistor and a voltage drop across the resistor of 6 volts, the resistance is $\frac{6}{10}$ or 0.6 ohms. The wattage rating of the resistor should be at least the current multiplied by itself times the resistance or a value of $10 \times 10 \times 0.6$ equals 60 watts.

Charger-Powered Voltage Regulator Test Set

To properly and quickly adjust and check automatic voltage regulators in battery chargers, a variable D.C. voltage and an accurate D.C. voltmeter are needed. A test set using voltage from the battery charger, instead of an external source of voltage, is shown in Figure 1-6.

In Figure 1-6a, potentiometer R2 is a 1K ohm, 10 watt wire wound unit. Terminals T1, T2 and T3 are small alligator clips attached to leads about 2 feet long. F is a $\frac{1}{10}$ ampere fuse. V is a D.C. voltmeter having a range suitable for the voltage of the battery, wired as shown.

To use the test set, disconnect the positive lead from the positive terminal of the voltage regulator and connect it to clip T1. Then, connect clip T3 to the positive terminal of the voltage regulator. Connect clip T2 to the negative lead and terminal, which is still connected to the voltage regulator. Connect the charger, turn it on and adjust potentiometer R2 until the correct cut-off voltage is reached. Readjust the voltage regulator if necessary.

If the battery voltage is not high enough to reach the cut-off point, the voltage of the charger at clips T1 and T2 can be increased above the battery voltage to speed up the adjustment by inserting a resistance R1 Figure 1-6b between the positive terminal of the battery BAT and positive battery clip CH of the charger. Resistor R1 can be an old steel spring having a wire gauge of No. 14 or larger depending on the current, $\frac{3}{4}''$ in diameter and 2 feet long, or a series of old tester discharge resistors. Better yet is the resistor board shown in Figure 1-4 and 1-5. Use a resistance value that will allow the charger voltage to rise slightly above the desired cut-off voltage of the regulator.

For further details on using this test set and adjusting voltage regulators, refer to Chapter 17, Section III.

Self-Contained, Battery-Powered Voltage Regulator Test Set

A self-contained, battery-powered voltage regulator test set that is ready-built and available from electrical supply houses is shown schematically in Figure 1-7.

V is a D.C. voltmeter with a full scale of 50 volts and a resistance of 50K ohms or 1K ohms per volt. Switch SW1 is a N.O. momentary contact switch that is closed for "test" and open for use as a separate voltmeter, BAT is a 67.5 volt "B" battery. R1 is a 10K ohm potentiometer and R2 is a 3K ohm resistor. The tester also has a built-in thermometer calibrated in cut-off voltage.

The voltmeter on this test set, due to the high potentiometer resist-

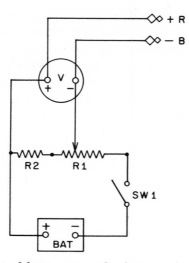

Figure 1–7   Self-contained battery-powered voltage regulator test set

ance, will drop suddenly when the zener diode voltage of the voltage regulator is reached. If it drops below the cut-on voltage of the regulator, increase the voltage by turning potentiometer R1 clockwise until the cut-off voltage is again reached. Then, turn the voltage down slowly to check the cut-on voltage, if so equipped, of the voltage regulator.

In the absence of more specific voltage regulator settings, use Table 1-1 as a guide. When the gassing point of the battery is reached during charging, the battery terminal voltage at 77 deg. F should be about as shown in Table 1-1.

Voltage Regulator Cut-Off Voltage for Various Batteries

12 Volt	(6 cell)	14.2 Volts	30 Volts	(15 cell)	35.6 Volts	
18 Volt	(9 cell)	21.3 Volts	32 Volts	(16 cell)	38.0 Volts	
24 Volt	(12 cell)	28.4 Volts	36 Volts	(18 cell)	42.6 Volts	Table 1-1

A.C. Powered Voltage Regulator Test Set

An A.C. powered voltage regulator test set is shown in Figure 1-8. Transformer T1 can be a 120 or 240 volt section of a small transformer with a center tap so it acts as an auto-transformer. Also either the primary or secondary section of an old radio power transformer can be used if it is center tapped. Diodes D1 and D2 are silicon diodes rated at 1 to 2 ampere at 200-400 PIV. The rectifier output must be filtered; otherwise, the peak rectified voltage would give a false trigger voltage. Normally, with the regulator connected to the charger, the battery acts as a large capacitor holding down most of these voltage peaks. Choke CH1 is a 30 henry choke. Capacitors C1 and C2 can be 8-16 mfd, 200 DCWV electrolytic capacitors. R1 is a 1K ohm, 10 watt wire wound potentiometer. F is a ¼ ampere fuse. Meter V is a 0-50 volt D.C. voltmeter. This circuit gives about 0-70 volts D.C. output. The auto-transformer can be a shock hazard. An isolation transformer with primary and secondary windings would be preferable for this reason if it has a 120 volt center-tapped secondary (60 volts either side of center tap).

Another method would be to use a bridge circuit in the rectifier section with an input voltage of 60 volts from a Variac or other variable voltage source.

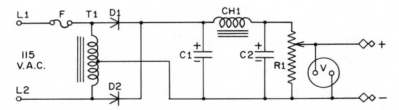

Figure 1–8   AC powered regulator test set

When using any test set that has an internal battery or a filtered rectifier circuit supplying the voltage, such as is shown in Figures 1-7 and 1-8, it is necessary to make a final adjustment after connecting the regulator to the charger because the charger output is not completely filtered except for battery voltage. Peaks of rectified voltage will trigger the regulator at a voltmeter reading below that set at a steady D.C. It will be necessary to increase the cut-off voltage adjustment. The cut-on voltage, however, should not need readjustment because the charger is off, and the battery alone is supplying the voltage.

By using the test set shown in Figure 1-6, this final adjustment is not usually necessary. However, it should be checked when the voltage regulator is reconnected to the charger output for normal operation.

Converter To Change Single-Phase To Three-Phase Power

If shop repair of large industrial chargers is attempted and a three phase power supply is not available, a relatively inexpensive converter can be built using any three phase electric motor in good condition, as shown in Figure 1-9, providing 230 volt single phase power is available. This converter will handle all 230 volt, three phase chargers and motor-generator sets, but not 440 volts, three phase. This could be done however, by using a 10-15 KVA single phase transformer to step up the voltage from 230 to 440 volts and by adapting the motor to 440 volt operation.

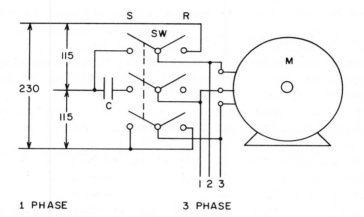

Figure 1–9   Single-phase to three-phase power converter

In Figure 1-9 the motor M is a 230 volt, three phase induction motor rated at 10-15 H.P.. Switch SW is a 3PDT knife switch with an inter-lock to prevent setting it to the run R position until after it has been set to the start S position. Switch SW can be made up of two separate 3PST switches placed side by side with the handles facing each other, thus placing one switch upside down. Bolt the two handles together and the result is a 3PDT switch, but without an inter-lock.

Motor M is started on 115 volts single phase using a 500-700 mfd. 120 volt electrolytic motor-starting capacitor C. As the motor comes up to full speed, throw switch SW to the run R position, which connects the single-phase 230 volts across one phase of the motor M.

A three-phase motor running on single-phase voltage generates a three phase voltage at the terminals of the motor, due to the magnetic induction of the turning rotor in the A.C. field. Connect the three motor leads 1, 2 and 3 to an outlet or a disconnect switch for three-phase testing.

This converter can be made self-starting by using a motor starting relay, which is available from electric motor parts supply houses. Wiring instructions are usually included with the relay for this type of converter.

Such a motor makes an ideal heavy duty grinder, by belt-driving an arbor holding a 10 inch blacksmith grinding wheel. There is enough starting torque in the motor to bring the wheel up to speed, and the inertia of the wheel acts as a flywheel to assist the motor during sudden high current loads, such as starting a three-phase motor-generator set.

Precision D.C. Calibrating Voltmeter

Anyone engaged in the calibration of battery testers, and adjustment of voltage regulators needs an accurate D.C. voltmeter having ½% guaranteed accuracy, or better.

There are no standard meters on the market that have suitable ranges and accuracy at a reasonable cost to properly calibrate and adjust battery service equipment.

The author has written the specifications, and been instrumental in having developed, an ideal D.C. voltmeter designed specifically for the special needs of the battery service equipment industry.

Figure 1-10 shows the scale of a three range D.C. voltmeter designed for those who are concerned only with vehicle starting batteries having voltages of 6, 12, and 24 volts.

Figure 1-11 shows the scale of a two-arc D.C. voltmeter having four basic ranges for 6, 12, 24 and 36 volt batteries, covering all vehicle starting batteries plus golf cart and other 36 volt batteries. The range selector, covering these four ranges is shown in Figure 1-11a. Figure 1-11b shows two additional ranges of 8×10 (40 to 80 volts read on the 4-8 scale and multiplied by 10), and 16×10 (80 to 160 volts, read on the 8-16 scale and multiplied by 10), for those who work with all battery voltages likely to be encountered, or anticipated in the industrial and electric vehicle field.

Electric meters having an accuracy of 2% and above are manufactured using pre-printed dials, which are all the same. Precision meters of ½% accuracy, or better, require individual dials and calibration scales, drawn

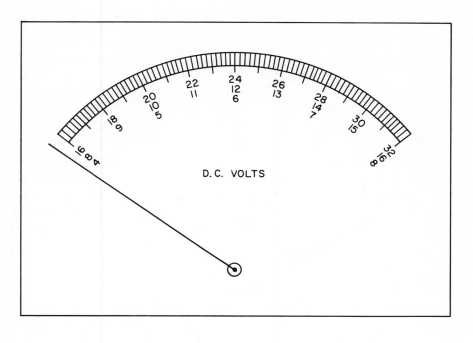

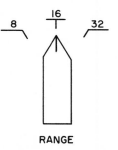

RANGE

Figure 1–10 Precision DC calibration voltmeter one-arc scale

by hand. The cost to manufacture these meters to specifications is about the same as that of a comparable standard scale meter listed in meter catalogs.

Until these meters are available on the meter market, they can be made to order at a cost comparable to that of a less accurate jewel type meter.

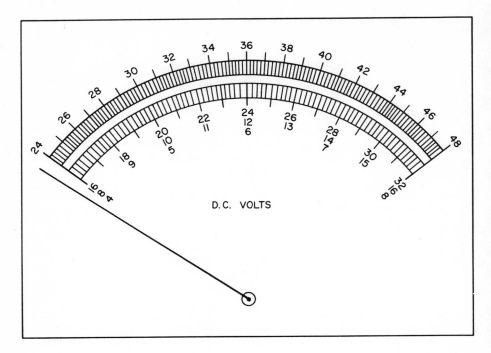

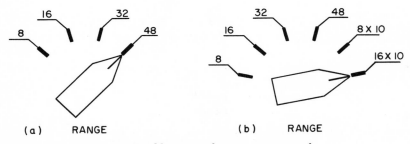

Figure 1–11 Precision DC calibration voltmeter two-arc scale

Source 14 can furnish these meters shown in Figures 1-10, 1-11a, 1-11b and Table 1-2. Using only reliable, blank dial meter movements from such manufacturers as General Electric, Westinghouse, Weston, and others, they are equipped to make up accurately calibrated dials. Since they are familiar with the specifications, and they can quote a price if the Figure number and the preferred meter size are given. For example: when writing for a price, state, "Please quote price and delivery on the Cantonwine meter, specifications Table 1-2, drawing Figure 1-11b in 6½" meter size."

Specifications for Precision D.C. Calibrating Voltmeter for Battery Service Equipment

			Drawing Number		
Specifications			*Fig. 1-10*	*Fig. 1-11a*	*Fig. 1-11b*
Minimum Meter Size			4½"	5½"	5½"
Recommended Meter Size			5½-6½"	6½-7½"	6½-7½"
Number of Arcs			1	2	2
Number of Ranges			3	4	6
Nominal Battery Volts and			6	6	6
Mid-Scale Volts			12	12	12
			24	24	24
				36	36
					60
					120
Scale	*V./Div.*	*V. @ ½% ±*			
4-8	.05	.04	X	X	X
8-16	.10	.08	X	X	X
16-32	.20	.16	X	X	X
24-48	.20	.24	—	X	X
8X10	.50	.40	—	—	X
16X10	1.00	.80	—	—	X
D.C. Voltmeter			X	X	X
±½% Accuracy			X	X	X
Taut-Band Suspension			X	X	X
Mirror Scale			X	X	X
Highly Damped Movement			X	X	X
Res. 100-1000 Ohms/Volt			X	X	X

Table 1-2

The price depends on the meter size, and whether it has one or two scale arcs. A meter with the additional ranges shown in Figure 1-11b (8×10 and 16×10) is only slightly more because only two precision multiplier resistors and two extra positions on the selector switch must be added.

The taut-band suspension meter is superior to and costs less than the jewel type because there are no bearings nor hair spring to cause friction. They are more rugged and reliable over a long period of time and readings are easily and accurately taken.

SPECIAL TOOLS REQUIRED 2

In addition to the test equipment discussed in Chapter 1, there are some tools and special equipment that will be helpful in servicing and repairing battery chargers. Some of these items are not absolutely necessary to do the job, but you will find that the work is done much more easily, neater, and better with them.

Air Compressor

A small air compressor is useful for blowing accumulated dust out of chargers, drawn in by air circulation.

Compressed air for spray painting is not necessary because paint and insulating varnish are available in pressurized cans.

Welding and Brazing Equipment

Welding and brazing equipment is seldom needed, but in any mechanical or electrical repair shop there are occasions when it will solve problems. All necessary arc welding, brazing, and heavy soldering can be done with the battery charger load tester and D.C. welder shown in Figure 1-3. Use long cables with a shield between the welding area and the batteries since the gas from charging batteries is explosive. A small 100 ampere A.C. welder or an oxy-acetylene torch can serve the purpose equally well.

Hand Tools

Every mechanical or electrical repair shop needs a good selection of hand tools, such as pliers, screwdrivers, wrenches. If necessary, start with a few tools and add other tools as you need them. A large variety of hand tools is necessary since many different kinds of fasteners are used in battery charger construction. Some tools may not be absolutely essential, but any tool that will save time or do a better job is a good investment.

Work Platform

A very handy piece of equipment, especially for the larger chargers, is a work platform on casters. This can be made from wood, or a wooden box with casters can be used. The platform should be approximately 24 inches high, with the top platform approximately 22 inches square. A shelf should be provided underneath for storing the cables, case panels, and large parts. A lip should be left on the edge of the top as a convenient place to secure the battery clamps. Make several of these platforms to use for rush jobs, and jobs being held for parts. Also, it allows you to work on a second charger while one is on test or running through a timing cycle. The height of the platform allows you to sit down to make repairs, tests, and adjustments. The casters permit the charger to be turned around or rolled to any part of the shop.

Label Making Machine

A label making machine is very useful in servicing battery chargers or building test equipment if the various controls need new markings. Many times the old markings cannot be read or are missing entirely or else new instructions must be added. The embossed tape labels add a professional look to the finished appearance of the job. Panel decals are available from many sources and they are easily applied by wetting them and laying them on the cleaned panel area.

Special Tools

Other special tools and equipment are shown in other chapters of Section III, such as, rectifier assembly tools (Chapter 1), diode tools (Chapter 2), solenoid repair tools (Chapter 12), and battery clamp repair tools (Chapter 19).

SECTION III

1
SELENIUM RECTIFIERS

Assemblies

This chapter discusses the various possible assembly combinations of selenium rectifiers.

Selenium rectifiers are constantly being improved for better conductivity, larger current carrying capacity, and voltage rating. In the absence of more specific data, assume that each selenium plate has a PIV rating of at least 50 volts, and may be used on 6 and 12 volt chargers. However, two rectifiers should be used in series for 24 volt chargers, and three rectifiers should be used in series for 36 volt chargers. The area in square inches (length × width) should be the same or greater than the original.

Selenium rectifiers are composed of the basic selenium cell or plate consisting of a selenium covered plate with conductive coating, a contact plate or washer (contacts of the spring type require a precision washer to limit the amount of pressure), and the necessary hardware.

The various arrangements likely to be encountered are shown in Figures 1-4 to 1-15, and may be obtained from manufacturers as completely assembled and tested units, by model number and description or part number. The arrangements shown are in schematic form. However, the same type and size of unit in different brands and models may have the connections brought out in different positions.

Completely assembled rectifier units can be supplied for various charger models as well as a variety of sizes of selenium plates and all necessary hardware to build or repair almost any conceivable type and size of selenium rectifier for any brand or model of battery charger. The basic selenium cell is assembled with the spring contact plate, or collector disc, sealing disc, and precision spacing washer held in place by an adhesive label. This assembly is called a "building block". Detailed assembly instructions for all types of rectifiers are available from many manufacturers upon request.

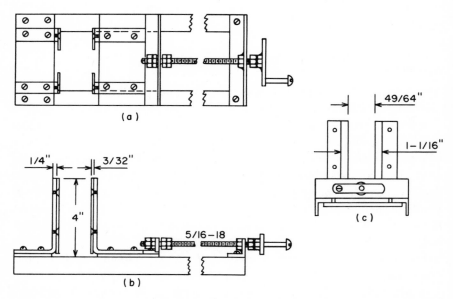

Figure 1–1   Selenium rectifier assembly vise-top view (a), side view (b), and right end view (c)

Assembly Tools

A clamping device, or vise, is necessary to assemble rectifier units from separate parts, to save time, and to prevent damaging the parts during assembly. Such a vise can be built by following the detailed drawings in Figure 1-1 and Figure 1-2. The important dimensions are shown. The vise jaws must be thin enough to allow tightening the last steel washer over the $\frac{1}{8}$ inch thick insulating washer. Therefore, a $\frac{3}{32}$ inch plate is screwed, using flat head countersunk screws, to the

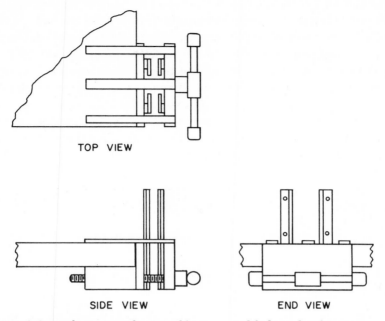

TOP VIEW

SIDE VIEW   END VIEW

Figure 1–2   Selenium rectifier assembly vise—modified woodworking vise

heavier jaws, which are $\frac{3}{16}$ inch to $\frac{1}{4}$ inch thick for added strength.

The vise must have two identical jaws 4 inches high, one stationary and the other movable over a range of 0-12 inches, that stay parallel to each other.

A pilot tool aids in assembling the rectifier stack. Details for making a pilot tool are shown in Figure 1-3. A 3 inch length of phenolic tubing having a bore of $\frac{5}{16}$ inch and an OD of $\frac{1}{2}$ inch (same as used in the rectifier stacks), is cemented on the end of a threaded rod $\frac{5}{16}$ inch-18 and 11 to 13 inches long as shown in Figure 1-3a. Use good cement such as the two-tube type of epoxy glue. After the glue has set, turn on a lathe, or grind to a point as shown in dotted lines in Figure 1-3a until the tool appears finished as in Figure 1-3b.

To use the pilot tube, slip a 10 inch length of phenolic tubing over the threaded rod. The rectifier stack is then assembled on this tube and placed in the vise jaws. Place the assembly in the vise with the bottom of all the rectifier plates resting on the bed of the vise. Center the tubing in the vise so the final insulating end washers will fit between the two thin lips on both jaws. If the collector disc is facing the vise jaws, such as for a half-wave rectifier or for a negative output full-wave rectifier, be sure a terminal strip is installed or the sealing disc will be damaged.

Crank the movable vise jaws against the rectifier stack and clamp into position temporarily until all terminal straps have been turned into the proper position. Then clamp tightly and, at the same time, wiggle and turn the pilot tool, keeping it loose enough so that it can be removed later. Now, mark the length of the tube required for the rectifier stack, which should be about $\frac{1}{16}$ inch shorter so it will not "dead end" on

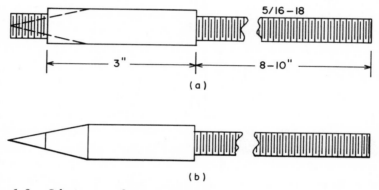

Figure 1–3 Selenium rectifier assembly pilot tool

the end of the tube when bolted together. If an exact length is not available, cut the exact length from a longer piece. Slip the tubing on the pilot tool, and insert it into the rectifier stack. The clamping bolt is installed with steel washers on each end and tightened. If the terminal straps can be moved, the rectifier stack is not tight enough, or the tube is too long and has come in contact with the steel washers, which does not permit a complete clamping of the stack.

If it is necessary to make an adjustment of the terminal straps after the stack has been removed from the vise, loosen the nut only enough to readjust the straps, then retighten.

Test rectifier after assembly, and install. There should be no electrical connection between the plates or terminal straps and the mounting bolt.

A simple and effective rectifier clamping vise can be made from a fast-acting woodworking vise, as shown in Figure 1-2. A slight turn of the handle to the left permits the movable jaws to be pushed in or out and, by turning the handle to the right, the vise clamps tightly. The jaws are already drilled for mounting a wood facing. Remove the wood facing and install the adapter jaws discussed earlier in this chapter.

A bed, used to support the rectifier plates, should be made of two or three straps, with one end attached to the movable jaw of the vise, and the other end protruding through the stationary vise jaws flush with the bench top. They should be long enough to permit full opening of the jaws (about 12 inches). This permits assembly of most rectifier stacks.

Remove the long handle or replace it with a shorter one to avoid applying too much pressure. *Use only enough pressure to compress the springiness of the collector discs on the rectifier plates.*

Half-Wave, Parallel Selenium Rectifiers

Figure 1-4 shows half-wave rectifiers, singly and in parallel.

Figure 1-4a shows a half-wave single selenium cell or plate. As in all half-wave units, it may be a positive output or a negative output, depending on its connection in the circuit. When one side of the transformer secondary is connected to the anode, the rectified current flows to the cathode, and therefore has positive output. Remember that the current flows from the anode to the cathode. Likewise, if the transformer secondary is connected to the cathode C, the output at the anode A will be negative.

The schematic symbols, shown under each diagram, show an arrow pointing in the direction of current flow from the anode A to the cathode C.

Each plate of a selenium rectifier has a definite amperage rating, depending on the conductivity, size, and cooling method. It also has a definite safe RMS A.C. working voltage, which is usually about $\frac{1}{3}$ of the PIV rating.

To increase the total amperage rating of the rectifier, two or more plates are connected in parallel at the same voltage rating of a single plate. To increase the voltage rating of the rectifier, at the same current rating of a single plate, two or more plates are connected in series.

Figure 1-4b shows two plates in parallel to provide a half-wave recti-

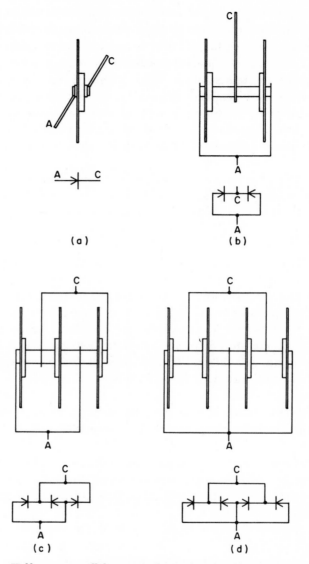

Figure 1–4   Half-wave, parallel-connected selenium rectifiers

fier having a 2 X ampere rating (twice the amperage rating of a single plate), and a voltage rating of 1 X (same voltage as a single plate).

Figure 1-4c has three plates in parallel to give a 3 X ampere rating and a 1 X voltage rating.

Figure 1-4d has four plates in parallel to give a 4 X ampere rating and a 1 X voltage rating.

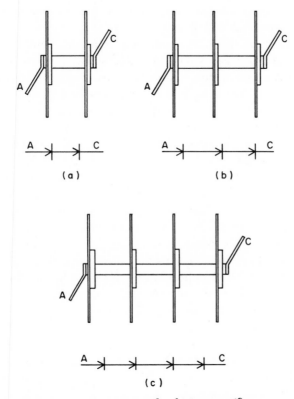

Figure   1–5   Half-wave, series-connected selenium rectifiers

Half-Wave, Series Selenium Rectifiers

Figure 1-5 shows half-wave rectifiers in series.

Figure 1-5a shows two half-wave plates connected in series to provide a 1 X ampere rating and a 2 X voltage rating.

Figure 1-5b shows three half-wave plates connected in series to provide a 1 X ampere rating and a 3 X voltage rating.

Figure 1-5c shows four half-wave plates connected in series to provide a 1 X ampere rating and a 4 X voltage rating.

Full-Wave, Positive Output Selenium Rectifiers, Single and Parallel Plates

Full-wave rectifiers using a center tapped transformer secondary have a 2 X ampere rating and a 1 X voltage rating.

Full-wave rectifiers may be assembled as a single unit, or they may be two separate half-wave plates mounted separately and connected in parallel to increase the ampere rating or connected in series to increase the voltage rating. They may also be connected for positive or negative output. This polarity choice is made by the manufacturer for the sole purpose of simplifying the wiring.

Figure 1-6a shows two separate half-wave plates connected for full-wave positive output to give a 2 X ampere rating and 1 X voltage rating. Usually these are mounted separately, one on either side of the case, which also acts as a heat sink. They must be insulated from the case.

The full-wave rectifier shown in Figure 1-6b is the same electrically as that shown in Figure 1-6a, but it is assembled as one unit, and is usually used on fan–cooled fast chargers.

The terminals on all full-wave rectifiers are identified by a yellow dot for the A.C. connection and either a red dot or stripe for the D.C. positive output, or a black dot for negative D.C. output.

Figure 1-6c has two sets of two parallel plates connected in a full-wave circuit, having positive output. This circuit gives a 6 X ampere rating, and a 1 X voltage rating.

Figure 1-6d has two sets of three parallel plates connected in a full-wave circuit having positive output. This arrangement has a 4 X ampere rating, and a 1 X voltage rating.

Full-Wave Negative Output Selenium Rectifiers, Single and Parallel Plates

The full-wave circuits shown in Figures 1-7a-d are the same as those shown in Figures 1-6a-d except that the plates are reversed to give a negative output.

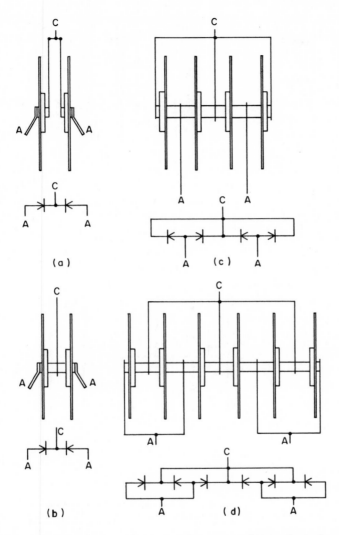

Figure 1–6   Full-wave, positive output selenium rectifiers with single and parallel plates

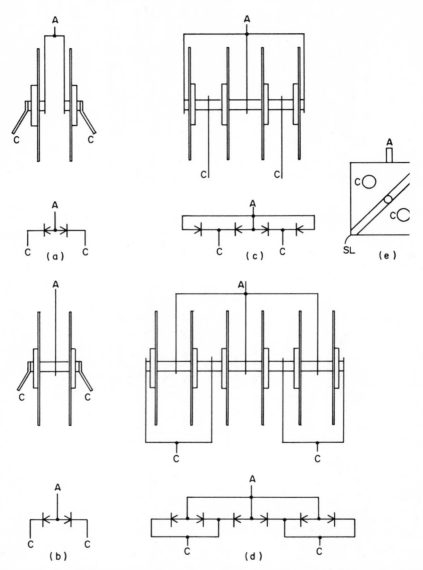

Figure 1-7 Full-wave, negative output selenium rectifiers with single and parallel plates

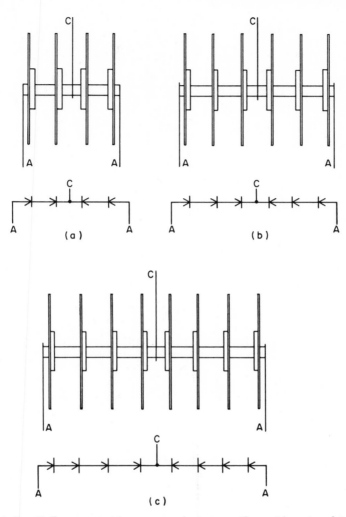

Figure 1–8   Full-wave, positive output selenium rectifiers with series plates

Figure 1-7e shows a special full-wave rectifier made from a single selenium plate. A slot SL is milled diagonally across the plate, cutting down to the metal base plate, leaving two separate selenium cells, for a negative output only. Two small contact buttons (C) forming the two cathode connections are held in position, under pressure, by an insulated bar. This type is limited to small battery chargers of 4-6 ampere rating.

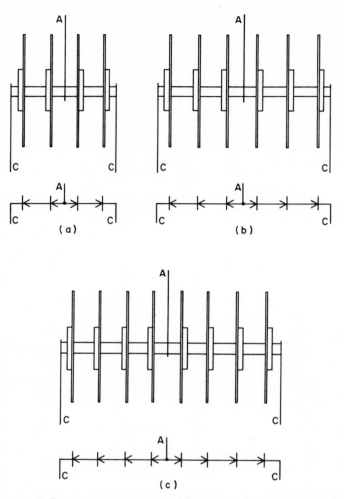

Figure 1–9   Full-wave, negative output selenium rectifiers with series plates

Full-Wave Positive Output Selenium Rectifiers, Series Plates

Figure 1-8 shows full-wave, series connected plates for a positive output.

Figure 1-8a shows 2 sets of 2 plates in series to provide a 2 X ampere rating, and a 2 X voltage rating.

Figure 1-8b shows 2 sets of 3 plates in series to provide a 2 X ampere rating, and a 3 X voltage rating.

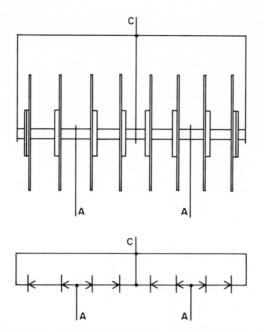

Figure 1–10 Full-wave, positive output selenium rectifier with series-parallel plates

Figure 1-8c shows 2 sets of 4 plates in series, providing a 2 X ampere rating, and a 4 X voltage rating.

Full-Wave, Negative Output Selenium Rectifiers, Series Plates

Figures 1-9a-c are similar to those shown in Figures 1-8a-c, except that the plates are reversed to provide a negative output.

Full-Wave, Positive Output Selenium Rectifier, Series-Parallel Plates

To increase both current capacity and voltage rating, the plates are connected in series and parallel, as shown in Figure 1-10, for a positive output. This provides a 4 X ampere rating, and a 2 X voltage rating.

Full-Wave, Negative Output Selenium Rectifier, Series-Parallel Plates

Figure 1-11 is similar to Figure 1-10 except that the plates are reversed to provide a negative output.

Full-Wave Bridge Selenium Rectifiers, Single and Series Plates

The full-wave bridge circuit does not require a center tap on the transformer secondary, as do other full-wave circuits.

Figure 1-12a shows a 4-plate bridge rectifier. The A.C. input, or transformer, terminals are identified by a stripe or dot of yellow paint,

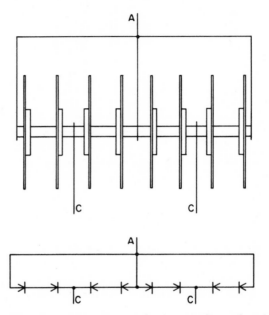

Figure 1–11 Full-wave, negative output selenium rectifier with series-parallel plates

the positive D.C. output terminal is marked in red, and the negative terminal in black or is plain. This rectifier may be assembled in one complete unit, or it may be two separate units as shown in Figure 1-12b. The upper rectifier is a full-wave positive output unit, and is shown in the right portion of the symbol. The lower rectifier is a full-wave negative output unit, and is shown in the left portion of the drawing. The two A.C. terminals are connected together and go to the transformer secondary. This arrangement has a 2 X ampere rating, and a 1 X voltage rating. Figure 1-13 has a 2 X ampere and 2 X voltage rating.

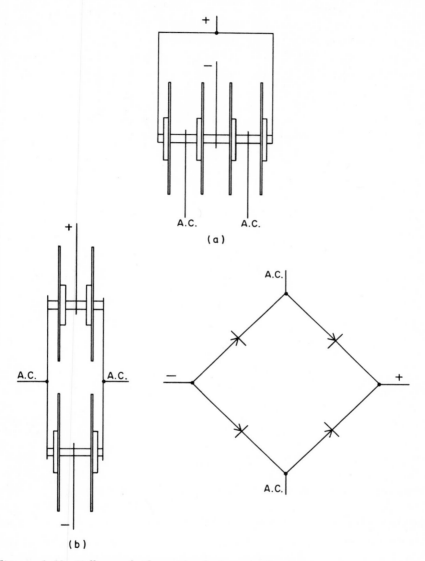

Figure 1–12 Full-wave bridge with selenium rectifiers

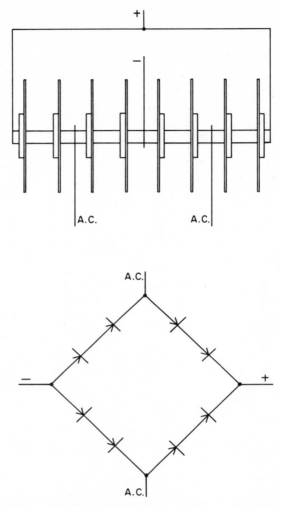

Figure 1–13 Full-wave bridge with selenium rectifiers, series plates

Full-Wave Bridge Selenium Rectifier, Parallel Plates

To increase the amperage rating of the full-wave bridge circuit, at the same voltage rating, two or more plates are connected in parallel as shown in Figure 1-14. This 8-plate rectifier provides a 4 X ampere rating and a 1 X voltage rating.

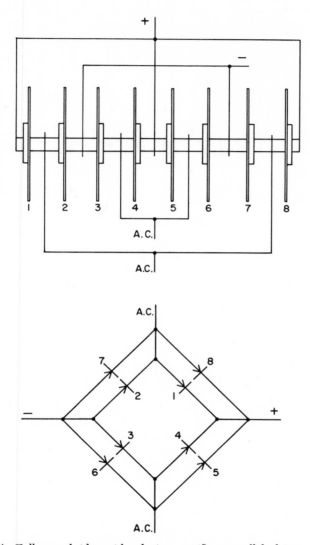

Figure 1-14 Full-wave bridge with selenium rectifiers, parallel plates

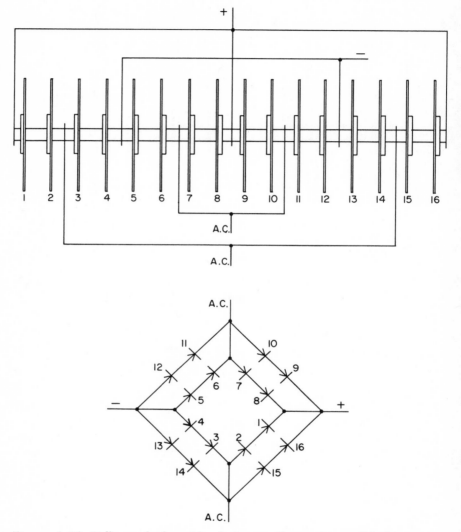

Figure 1–15 Full-wave bridge with selenium rectifiers, series-parallel plates

Full-Wave Bridge Selenium Rectifier, Series-Parallel Plates

To increase both the ampere and voltage ratings, two or more plates are connected in series, and two or more plates are connected in parallel, forming a series-parallel arrangement as shown in Figure 1-15. This 16-plate rectifier provides a 4 X ampere rating, and a 2 X voltage rating.

In a long stack such as this, all terminals must be the same length, or terminate in an extra-heavy bus-bar to equalize the load. Otherwise, the plates having the shorter leads would be overloaded, and overheat.

2
SILICON DIODES

Silicon diodes are made in shapes that are different from those of the selenium type plates. Silicon diodes have built-in series arrangements for various peak voltage ratings, and their physical size usually indicates a relative amperage rating. The silicon diode must be mounted on an adequate heat sink, either by pressing into a reamed hole or by stud mounting. Smaller silicon diodes are mounted by leads, usually without any other heat sink than that provided by the soldered connection.

Often, silicon diodes are mounted on the heat sink in series and/or parallel arrangements. The primary reason is to simplify stocking several different ratings. For example, a manufacturer uses a certain low cost silicon diode in their fastest selling popular charger. When they make small quantities of higher output chargers, they use their standard diode by connecting two or more in either a parallel or a series arrangement. The overall result is economy. They could use a more expensive single diode to do the same job.

All connections shown for selenium rectifiers in Chapter 1, Figures 1-4 to 1-15 can be duplicated with silicon diodes, and the symbols are the same.

Silicon Diode Polarities

Selenium rectifier plates, or cells, are always of one polarity; that is, the bare metal plate is always the anode, and the selenium coated side with the collector disc is always the cathode or positive output. Of course,

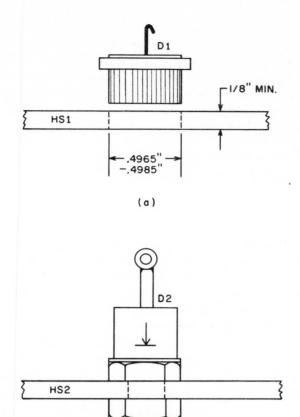

Figure 2–1 Silicon diode heat sink mountings

the polarity of the output can be reversed by applying the A.C. to the cathode for negative output at the anode.

Silicon diodes may also be reversed, but the base or mounting stud may be manufactured for either a positive output or a negative output. Silicon diode polarity is identified by color, letter or an arrow. A diode having the base with negative output will have the type number printed in black, or a suffix letter following the type number, such as N for negative output at the base, or an R (reverse polarity). A diode having the base with positive output will have the type number printed in red or a suffix letter following the type number, such as P for positive output, or no letter at all, to indicate positive or normal output. For example, a Motorola type 1N1183 is a positive output at the base, but a Motorola type 1N1183R would have a negative output at the base. An arrow is often used in addition to the other designations. The arrow always points from the anode to the cathode, or the direction of current flow.

Silicon Diode Heat Sink Mountings

Figure 2-1 shows the two most common methods of mounting silicon diodes on the heat sink.

Figure 2-1a shows the press-in type silicon diode D1, and the heat sink HS1. The heat sink should be at least $\frac{1}{8}$ inch thick for the press-in type of silicon diode. To make heat sinks, or to drill new holes, use a $3\frac{1}{64}$ inch drill, and ream out the hole to 0.4965-0.4985-inch. An expansion reamer can be adjusted to give this dimension, or the trial and error method can be used until a good press fit is obtained. These diodes must be pressed in, using the tools shown in Figure 2-2. They should not under any circumstances be driven in with a hammer because certain damage to the diode will result. If an arbor press is not available, use a vise, and make the tools as short as possible to fit between the jaws of the vise.

Figure 2-1b shows the stud-mounted silicon diode D2, and the heat sink HS2. The polarity is indicated by an arrow. The heat sink for the stud-mounted type is usually less than $\frac{1}{8}$ inch thick, but the press-in type diode requires a heat sink $\frac{1}{8}$ inch thick or thicker. The drilled holes should be free of burrs, so that the base of the diode makes good contact with the heat sink. The nut should be tight, but not too tight. Use a torque wrench to avoid exceeding the manufacturer's recommended torques.

Some stud-mounted diodes have a right-hand thread for negative output diodes, and a left-hand thread for a positive output diode. Some diodes may be found to be the reverse of this.

To make good electrical and heat conductivity contact, some diodes are soldered to a copper heat sink with a low temperature solder. The chance of diode failure can be reduced if a good heat conducting compound is used on all contact areas every time a diode is replaced. A good compound to use is Dow Corning Type 340 or equivalent, silicon grease or heat sink compound.

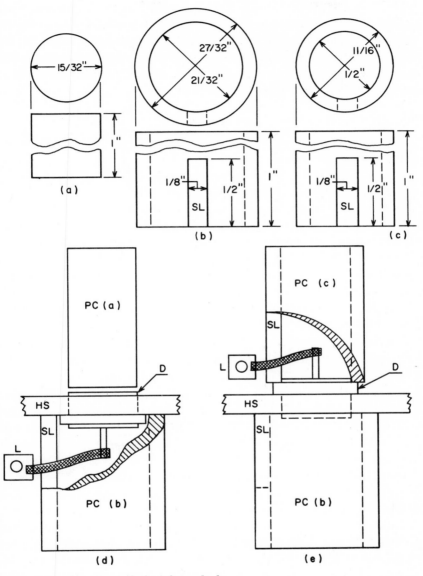

Figure 2–2   Pressing tools for silicon diodes

Press–in Tools For Silicon Diodes

The tools shown in Figure 2-2 are made from readily available material and are for removing and installing press-in type silicon diodes.

Piece a, Figure 2-2a is solid steel stock $^{15}\!/_{32}$ inch in diameter, and cut to 1-1½ inch length (a length of ½ inch rolled-thread bolt is about the right diameter). This is used to press out the diode.

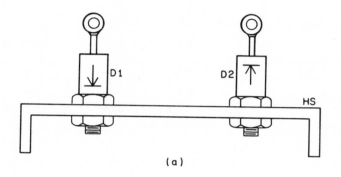

(a)

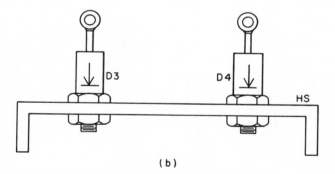

(b)

Figure 2–3   Series half-wave and full-wave silicon diodes

Piece b, Figure 2-2b, made by boring out a 1 inch long piece of ½ inch iron pipe, is used as a heat sink HS rest for pressing the diode D in or out.

Piece c, Figure 2-2c is made by boring out a 1 inch long piece of ⅜ inch iron pipe to an inside diameter of ½ inch.

Pieces b and c have a ⅛ inch wide by ½ inch long slot SL for diodes that have a large terminal on the end of a flexible lead L.

Figure 2-2d shows piece b used as a rest for the heat sink HS while piece a presses diode D out of heat sink HS. Piece b is shown cut away to give a view of the diode D lead L protruding through the slot SL.

Figure 2-2e shows piece b used as a rest for the heat sink HS while piece c presses diode D into heat sink HS. Piece c is shown cut away to give a view of the diode D lead L protruding through the slot SL.

Half-Wave Series, and Full-Wave Silicon Diodes

Diodes shown in Figures 2-3a and 2-3b have the same physical appearance, but Figure 2-3a shows a positive base diode D1 and a negative base diode D2 mounted on a heat sink HS and series connected as a half-wave rectifier commonly used in multiple battery chargers.

Figure 2-3b shows two diodes D3 and D4 having the same polarity connected as a full-wave rectifier. If the two diodes have a positive base, as shown, the D.C. output will be positive. If the two diodes are of the negative base type with the arrows reversed, then the D.C. output will be negative.

Half-Wave Parallel-Series, and Full-Wave Parallel Silicon Diodes

Diodes shown in Figures 2-4a and 2-4b have the same physical appearance, but Figure 2-4a shows two positive base diodes D1 and D2 connected in parallel, and two negative base diodes D3 and D4, connected in parallel to form a parallel-series half-wave rectifier, commonly used in multiple battery chargers. Note that this combination is designated parallel-series rather than series-parallel. A series-parallel connection requires that the elements, or diodes, be connected, first in series, and the series combinations placed in parallel with each other. And the parallel-series connection requires that the elements, or diodes, be connected, first in parallel with each other, and then the parallel combinations connected in series with each other. The over-all result is the same if all elements are the same.

Figure 2-4b shows two positive base diodes D5 and D6 connected in parallel, and two positive base diodes D7 and D8 connected in parallel forming a full-wave positive output rectifier. For negative base diodes with the arrow reversed, the D.C. output would be negative.

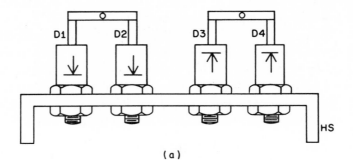

(a)

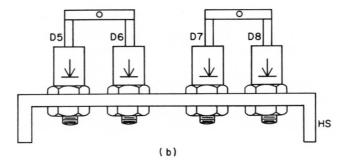

(b)

Figure 2–4   Half-wave parallel-series (a), and full-wave parallel (b) silicon diodes

Full-Wave Bridge Silicon Diodes

A full-wave bridge circuit is shown in Figure 2-5. This circuit requires two heat sinks HS, insulated from each other, one having a pair of positive base diodes D3 and D4, and the other having a pair of negative base diodes D1 and D2. One heat sink can be used to mount all four diodes, but each pair of negative or positive diodes must be insulated from the heat sink and the studs connected together by a jumper wire. The insulation does not conduct the heat as well as direct contact, so this method is usually limited to low current ratings.

Silicon diodes are available in one-piece bridge circuit construction, require a minimum of four connections, and use only one heat sink. The two yellow leads are the A.C. input, the black lead is negative output, and the red lead is the positive output.

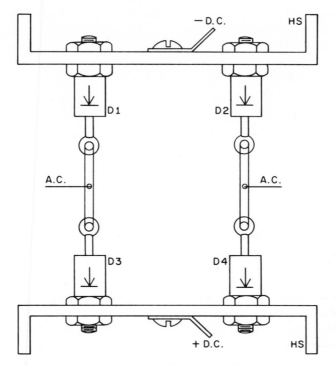

Figure 2–5   Full-wave bridge circuit with silicon diodes

Silicon Diode Ratings

Silicon diodes are available in PIV ratings of 50, 100, 200, 300, 400, 500 and higher. The higher the voltage rating, the greater the voltage drop. Often, a silicon diode rectifier is used as a replacement part for an original selenium rectifier. The reason they can have the same output as a selenium rectifier is that a replacement silicon diode with a 200-300 PIV rating has about the same voltage drop, or conductivity, as a 50 PIV selenium rectifier.

A rectifier diode should be able to withstand a voltage 2.5 times the normal working peak inverse voltage. This is a NEMA standard. This requirement provides a design margin for transient voltages arising from A.C. line surges. Therefore, the ratio of the diode PIV rating to the operating PIV should be at least 2.5 to 1. For example, when a diode is subjected to a peak inverse voltage of 20 (14.07 RMS × 1.414), the PIV rating should be at least 50 (20 × 2.5) volts. This safety factor assumes the use of transient protection, such as selenium surge protectors, capacitors or resistor-capacitor networks across the rectifiers.

Battery Isolation Circuits, Using Silicon Diodes

Any half-wave diode or rectifier, such as silicon, selenium, germanium and so forth, can be used to isolate certain electrical components. For example, a related application can be found where it is desirable to isolate two or more batteries from each other when on the same charging circuit.

Figure 2-6a shows two batteries in parallel across the charger, or generator G, with a diode D1 in series with battery BAT 1, permitting both batteries to take the charge. There will be a small voltage drop across diode D1 so that battery BAT 1 will probably charge at a slightly lower rate than battery BAT 2. On discharge, the LOAD 2 on battery BAT 2 will draw no current from BAT 1 because of the blocking action of diode D1. However, the LOAD 1 on BAT 1 could draw some current from BAT 2. This simple isolation circuit can be used in a common application, such as a vehicle and a trailer, each having its own battery. In this case, the vehicle battery would be BAT 1 and the trailer battery would be BAT 2. The trailer battery BAT 2 could be used and completely discharged without discharging the vehicle battery. When starting the vehicle on BAT 1, a small assist would be obtained from the trailer battery BAT 2. The diode D1 should be at least 100 amperes, 50 PIV and mounted on a large heat sink in a well-ventilated location in the air stream of the vehicle radiator fan away from the heat of the engine. The voltage regulator should be adjusted to give a compromise voltage on each battery due to the small voltage drop across diode D1. In other words, the cut-off voltage would be a little above normal on BAT 2, and a little below normal on BAT 1.

Two diodes D1 and D2 are shown in Figure 2-6b which isolate both batteries from each other. LOAD 1 will draw current only from BAT 1, and LOAD 2 will draw current only from BAT 2. The voltage regulator should be adjusted to give normal cut-off voltage across each battery, which should be the same for each battery, not at the generator output because of the small voltage drop across the diodes. The diodes D1 and D2 should be of the same rating, and can be mounted on the same com-

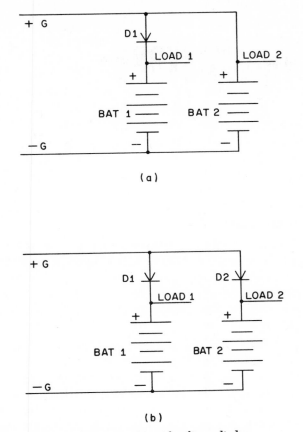

Figure 2–6 Battery isolation circuits with silicon diodes

mon heat sink if they have a negative base, since the bases are common. In this circuit, the diodes D1 and D2 can be rated to carry the charging current to each battery. A 35-50 ampere, 50 PIV unit should be ample for a 50 ampere alternator or generator total charging current, since each battery will share about one half of the output if equally discharged. Only two batteries are shown in Figure 2-6 but any number of batteries and diodes may be used.

Silicon diodes are best suited for this application because of their low voltage drop and long life.

TRANSISTORS ③

PNP Transistor Structure

Transistors used in battery charger alternator protectors (reverse polarity protectors) are heavy duty types that handle high currents. Smaller lower capacity transistors are found in other control circuits, such as voltage regulators in automatic chargers.

A transistor is similar to two silicon diodes connected back-to-back, as shown in Figure 3-1b.

A single silicon diode structure is shown in Figure 3-1a, where A is the anode and C is the cathode. The heart of a silicon diode is a small wafer of silicon having two kinds of impurities, one in the top half, or P electrode, and one in the bottom half, or N electrode. It might be compared to a blotter having red ink on one side and blue ink on the other side. Where the red and blue ink meet in the middle is the junction where P type of impurity meets the N type of impurity. This junction permits current to flow from the P region to the N region, but blocks the flow of current in the reverse direction.

Another layer of impurities can be added as shown in Figure 3-1b (where A are the anodes and C are the cathodes), forming a PNP sandwich. But notice that, no matter how we connect it, one junction would allow current flow from P to N, but the other junction blocks it.

If a lead is connected to the center layer N as shown in Figure 3-1c, the result is a transistor. With battery BAT 2 and resistor R2 connected as shown, there will be no current flow without battery BAT 1 and resistor R1 connected. Now when BAT 1 and resistor R1 are connected

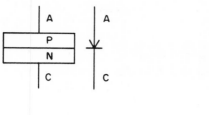

(a)

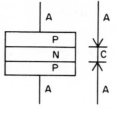

(b)

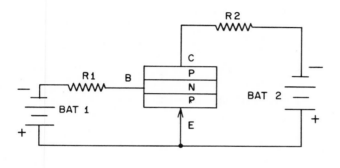

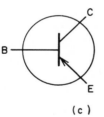

(c)

Figure 3–1 PNP transistor structure

as shown, current will flow through BAT 1, the lower P layer (emitter E), and through the center layer N (base B), and back to the BAT 1 through resistor R1. This current flow now permits current to flow from battery BAT 2 through the bottom layer P (emitter E), center layer N and top layer P (collector C), and back to the battery BAT 2 through resistor R2. If the polarity of the base current (BAT 1) is reversed, there will be no current flow in the collector-emitter circuit, the same as if there were no base voltage. The bottom diagram of Figure 3-1c shows the symbol of the PNP transistor. The arrow points in the direction of current flow.

The base current of a transistor need be only a small fraction (about 3%) of the collector current to maintain the flow of collector current. As the base current is increased by lowering the resistance of R1 or by

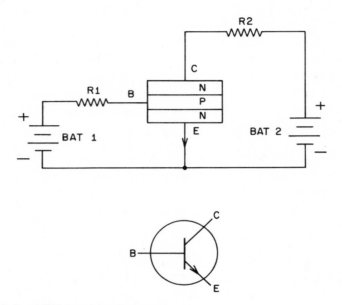

Figure   3–2   NPN transistor structure

increasing the battery BAT 1 voltage, the collector current will increase in direct proportion to the increase in the base current. This characteristic enables the transistor to amplify or multiply current signals, or to use small currents to operate heavier current solenoids, relays and other devices in battery chargers.

Figure 3-1c shows a PNP transistor, commonly found in battery charger circuits. However, both the PNP and NPN types of transistors are used. The only difference between the PNP and the NPN transistors is the polarity of the connections.

Transitors are similar in some ways to the three-element vacuum tube in which the grid voltage, at no current, controls the flow of current from the plate to the cathode, and the plate voltage changes. The vacuum tube is a voltage amplifying device, and the transistor is a current amplifying device that uses the base current to control the current flow through the collector and emitter.

NPN Transistor Structure

Figure 3-2 shows an NPN transistor having the layers stacked NPN, in opposite order to that of the PNP type in Figure 3-1c. Also, the battery polarities are reversed and the current flows in the opposite direction.

Transistor symbols for each type of transistor used in schematic wiring diagrams are shown in the lower drawings of Figure 3-1c and Figure 3-2. The arrow point is on the emitter and shows the direction of both the base and collector currents flowing through the emitter, just as with rectifier and diode symbols.

PNP and NPN Transistor Connections

The most popular of the many transistor case styles are shown in Figure 3-3. Note that the collector is connected to the metal case in most transistors, so it is essential that the case be insulated from the battery charger case, or mounted on a heat sink that is insulated. The type shown in Figure 3-3a may be mounted in a socket, or it may be soldered in the circuit. When soldering or unsoldering any small diode or transistor, use a heat-conducting clamp on the terminals between the solder joint and the unit to prevent damage from heat. A good clamp to use is a pick-up tool that is spring loaded to hold it closed; or use a pair of long nose pliers with a large rubber band around the handles to keep them closed.

Figure 3-3 illustrates both the PNP and the NPN transistors. The appearance and connections are the same, but the polarities are reversed internally. The transistors shown in Figures 3-3a and 3-3b are usually of the power output type, and those shown in Figures 3-3c and 3-3d are of the lower current capacity type.

Testing Transistors Using Polarity Tester

Transistors can be tested using the rectifier tester described in Section II, Chapter 1, Figure 1-2. Using prods PR1 and PR2, connect one of the prods to the base B of the transistor, as shown in Figure 3-4, and connect

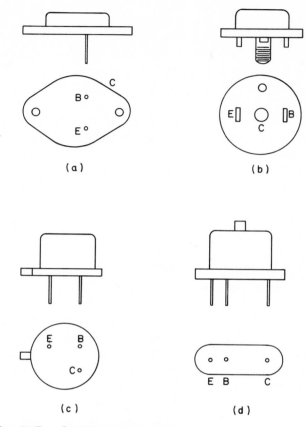

Figure 3–3   PNP and NPN transistor connections

the other prod first to the emitter E, and then to the collector C. If the
lamp connected first to the emitter E and then to the collector C glows,
and the base B lamp does not glow, and neither lamp glows when connected between collector C and emitter E, then the transistor is not
open, not shorted, has good conductivity, and is a PNP type as shown
in Figure 3-4a. If the base B lamp only glows when the other prod is
touched first to the emitter E and then to the collector C, and neither
lamp glows when connected between collector C and emitter E, then the
transistor is not open, not shorted, has good conductivity, and is a NPN

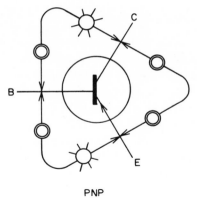

PNP

(a)

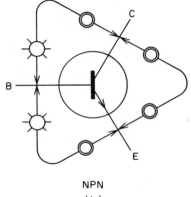

NPN

(b)

Figure 3-4   Transistor testing using a polarity tester

type as shown in Figure 3-4b. Neither lamp should glow when connected across the emitter E and collector C for either the PNP or the NPN type. If neither lamp glows when it should, the transistor is open. If both lamps glow at the same time, the transistor is shorted. In either case, the transistor should be replaced without further testing.

This test is enough in most cases, but power transistors may need a conductivity test, also.

Power Transistor Conductivity Test

Power transistors, as used in alternator protectors, can be tested for conductivity using prods PR3 and PR4 of the rectifier tester described in Section II, Chapter 1, Figure 1-2. Most of the smaller transistors used in voltage regulator circuits may also be tested for conductivity on this tester, but the test should be made as rapidly as possible to avoid over-heating. Usually, if transistors pass the test with prods PR1 and PR2, they will work satisfactorily in battery charger circuits.

When first testing with prods PR1 and PR2, determine the type and polarity of the transistor, that is, whether it is a PNP or NPN type, and make a diagram showing which terminal caused the lamp to glow. This terminal should be connected to the positive red lead PR3, and the other terminal should be connected to the negative black lead PR4. Always use the low range (under 5 amperes) for testing transistors. The current on this range is 100 MA.

For a PNP type transistor, as shown in Figure 3-5a, connect the black lead PR4 to the base B of the transistor, and the red lead PR3 to the collector C. The meter should read in "safe". Next, connect the red lead PR3 (with the black lead PR4 still on the base B) to the emitter E, and the meter should read in "safe".

For an NPN type transistor, as shown in Figure 3-5b, connect the red lead PR3 to the base B and leave it there while the black lead PR4 is connected, first to the emitter E, and then to the collector C. In each case the meter should read in "safe".

It is easy to remember the type designation by looking at the arrow in the symbol. If the arrow on E points to B, E is Positive, B is Negative and C is Positive, thus PNP. Likewise, if the arrow on E points away from the base B, then E is Negative, B is Positive and C is Negative, thus NPN.

Testing Transistors Using Ohmmeter or Test Lamp

Transistors can be tested by using an ohmmeter or a test lamp. The test lamp can be a #53 bulb, or other low current bulb, wired in series with a 6 or 12 volt battery. The polarities should be clearly identified. Ohmmeters are marked for polarity.

Figure 3-6 shows five separate tests for PNP transistors. To test NPN-type transistors, reverse the polarity of the ohmmeter or test lamp. Where

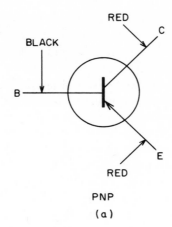

PNP

(a)

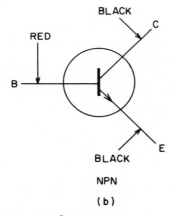

NPN

(b)

Figure 3–5 Power transistor conductivity test

positive is shown, connect the negative test lead, and where the negative is shown, connect the positive test lead. The results are the same for either type.

Figure 3-6a shows the ohmmeter, or test lamp, M with the positive lead connected to the emitter E and the negative lead connected to the collector C. The test lamp should not glow. The ohmmeter should read between 100-50,000 ohms. The actual value is not important.

Figure 3-6b shows the ohmmeter, or test lamp, M with the positive lead connected to the emitter E, the negative lead connected to the collector C, and a jumper J connected between the collector C and the base B. The test lamp should glow. The ohmmeter should read less than in Figure 3-6a and should be about 10 ohms or less.

Figure 3-6c shows the ohmmeter, or test lamp, M with the positive lead connected to the emitter E, the negative lead connected to the collector

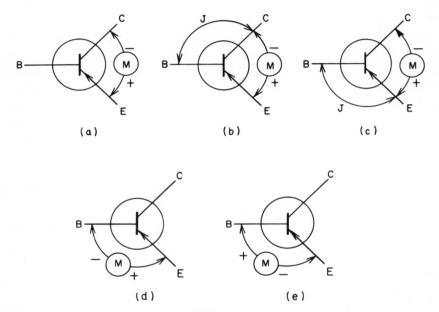

(a)   (b)   (c)

(d)   (e)

Figure 3–6   Transistor testing using an ohmmeter or test lamp

C, and a jumper J connected between the emitter E and the base B. The test lamp should not glow. The ohmmeter should show a resistance approximately 10 times higher than that shown in the Figure 3-6a test.

Figure 3-6d shows the ohmmeter, or test lamp, M with the positive lead connected to the emitter E and the negative lead connected to base B. The test lamp should glow. The ohmmeter should show a resistance slightly higher than that shown in the Figure 3-6b test (approximately 10-20 ohms).

Figure 3-6e shows the ohmmeter, or test lamp, M with the positive lead connected to the base B and the negative lead connected to the emitter E. The test lamp should not glow. The ohmmeter should read about the same or higher than that shown in the Figure 3-6c test.

The exact values of resistance are not important, but the ratios shown are. It is important to use the highest ohmmeter range possible to get a good reading. The resistance will vary on different ranges due to differences in the current through the meter and transistor on each range of the ohmmeter.

Figures 3-6b and 3-6c simulate the normal function of the transistor in battery charger circuits, particularly in the alternator protector circuit. Figure 3-6c shows the base with zero bias (a positive bias would have the same effect). The collector C to emitter E circuit has a high resistance and does not conduct. In Figure 3-6b, the base has a negative bias, the collector C to emitter E circuit has a very low resistance, and does conduct. The operating solenoid of the alternator protector in series with the collector C and emitter E will close when the base B is negative biased, but will open with either a positive bias or no bias on the base B relative to the common emitter E.

The characteristics and ratings of common transistors can be found in electronic catalogs or are available from the manufacturer.

4
SILICON CONTROLLED RECTIFIERS

Rectifier Identification

Silicon controlled rectifiers (SCRs), used in battery charger control circuits, such as alternator protectors, reverse polarity protectors, and voltage regulators, are identified in schematic wiring diagrams by the symbol shown in Figure 4-1d and by physical appearance. The most common types are shown in Figures 4-1a, 4-1b, and 4-1c. The three connections shown, anode A, cathode C, and control gate G input, are used with all SCRs.

Rectifier (SCR) Structure

The silicon controlled rectifier, shown in Figure 4-2a, is a PNPN diode with a third lead connected to the P layer next to the cathode C layer. The junction formed operates, and is tested, like a diode if the P layer is the anode and the N layer is the cathode. With a positive voltage on the anode and a negative voltage on the cathode, the SCR will conduct when enough positive voltage is applied to the control gate G input. This voltage will cause the SCR to conduct as long as there is enough current flowing between the anode A and the cathode C. To stop the SCR from conducting, disconnect the anode A, reduce the current level below the conduction point, or reverse the polarity of the current.

A simple PNPN SCR operates like a two-transistor circuit as shown in Figure 4-2b and Figure 4-2c. Transistor Q1 is an NPN type and transistor Q2 is a PNP type. These two transistors are connected in a regenerative feedback circuit. The output of transistor Q1 is connected to the input of

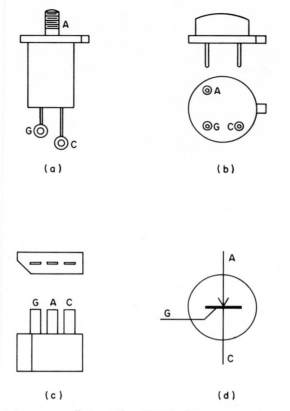

Figure 4–1 Silicon controlled rectifier (SCR) shapes, connections, and symbol

transistor Q2; and the output of transistor Q2 is fed back to the input of transistor Q1. With no bias voltage or a negative bias voltage on the control gate G input, there is no conduction between the anode A and the cathode C. However, when control gate G is biased positively, transistor Q1 conducts, providing a regenerative input to transistor Q2. Each transistor drives the other transistor into full conduction (saturation). Once conduction starts, all junctions are forward biased and the total voltage drop is the same as the voltage drop across a single PN junction or diode. Conduction will continue until the anode A to cathode C current is removed.

If the anode A to cathode C current is a steady D.C., the SCR will

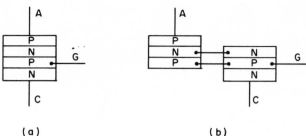

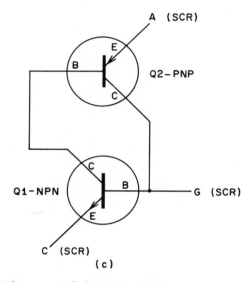

Figure 4–2   Silicon controlled rectifier (SCR) structure

conduct. However, if the anode A to cathode C current is an A.C. or pulsating D.C., the SCR will stop conducting when the current level goes below the conduction point. if there is enough positive bias voltage on the control gate G, the SCR will conduct again when the current level increases to the conduction point.

An SCR may be used as the main rectifier in a battery charger using the anode A and cathode C as a half wave rectifier. The charging current can be controlled by adjusting the bias voltage on control gate G input. If the gate G is biased negatively, such as, by connecting the battery in a reverse polarity, the SCR will not conduct.

Testing

An SCR may be tested by using a lamp and battery test set or an ohmmeter. The lamp used in the test set should have enough current capacity for the SCR being tested. The rectifier and transistor test set described in Section II, Chapter 1 may be used only for a polarity check of the SCR. Because of the pulsating D.C. current used in the test set, the SCR cannot be tested properly.

The polarity and conduction tests shown in Figure 4-3 are done using a lamp and battery test set. If an ohmmeter is used, resistance readings will vary from low (lamp will glow and SCR conducts) to high (lamp will not glow and SCR will not conduct).

POLARITY TEST. Connect the lamp and battery test set as shown in

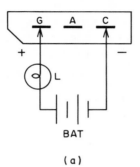

(a)

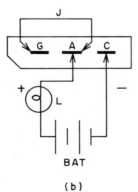

(b)

Figure 4–3   Silicon controlled rectifier (SCR) testing

Figure 4-3a. The lamp will glow if the SCR is good. Reverse the test set connections. The lamp should not glow. If it does, the SCR is shorted and must be replaced.

CONDUCTION TEST. Connect the lamp and battery test set as shown in Figure 4-3b. Do not connect jumper wire (J) yet. If the lamp glows, the SCR is shorted and must be replaced. If the SCR is good, momentarily touch the jumper wire between the gate G and the anode A until the lamp lights. The lamp should stay lit after removing the jumper wire. If the lamp does not stay lit, it may not pass enough current to keep the SCR conducting. Either use another lamp with a higher current capacity or use a 12-volt battery in place of the 6-volt battery. If using a No. 53 lamp (0.08 amp @ 6 v; 0.12 amp @ 12 v), try using a No. 1073 lamp (1.25 amps @ 6 v; 2.0 amps @ 12 v). Test the SCR only long enough to get a good indication. Be sure the SCR does not overheat.

If using an ohmmeter, the SCR should conduct on the lowest resistance range if the meter current is large enough. The SCR may conduct momentarily if the meter current is very close to the conduction level. Larger SCRs may not conduct as long as smaller SCRs. However, with enough current, any size SCR should conduct as long as the circuit is complete.

5
ZENER DIODES

Zener diodes are special in that they operate at reverse polarity. They act as a normal diode in the forward direction, with the positive voltage to the anode and negative voltage to the cathode, and can conduct heavier currents than a zener diode. They also act as a normal diode in the reverse direction, by having a high resistance up to the point of maximum peak reverse voltage, or zener voltage. The normal diode will break down and conduct heavily at this point and overheat rapidly or explode. The zener diode, however, is designed to conduct safely within certain limits and the voltage drop is fairly constant for any specific current within its operating range.

This constant voltage characteristic makes the zener diode ideal as a voltage regulator control or reference point. At the present time, zener diodes are available only in relatively low current ratings. Therefore, to handle larger currents for control circuits they are used with transistors, SCRs and relays or solenoids, acting only as a triggering device.

Zener Diode Shapes and Symbols

Figure 5-1 shows some common shapes of zener diodes. Figure 5-1a shows the higher wattage ratings of 10-50 watt size, Figure 5-1b shows the "top hat" type, Figure 5-1c shows the flat or circular type, Figure 5-1d and Figure 5-1e show two forms of the small tubular type, and Figure 5-1f and Figure 5-1g show two commonly used symbols. The polarity signs, by arrows, show the normal diode forward direction. The zener diode connection is the reverse of these, and is shown by a plus or minus sign, the polarity for zener operation.

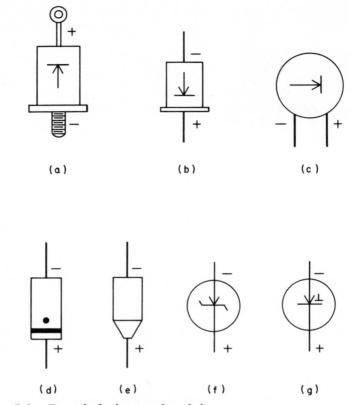

Figure 5–1   Zener diode shapes and symbols

Zener diodes are used in solid state automobile generator and alternator voltage regulators, and in battery chargers to operate voltage regulators. They are used in automatic chargers to maintain a constant battery voltage, to turn the charger off, or reduce its charging rate by a system of zener diodes, transistors, SCRs and relays.

Zener diodes may also be found in some newer expanded range D.C. voltmeters used on battery testers. For example, if an 8 volt zener diode is used in the voltmeter circuit, the meter pointer would stay at zero until 8 volts is reached. Then, it would show the voltage only from 8 to about 16 volts. With only 8 volts spread over the entire meter scale, the voltmeter can be read much more accurately than if the 16 volts were spread over the entire dial. In battery testers, readings outside of the 8-16 volt range, for example, are relatively unimportant.

Zener Diode Ratings

The zener diode is available in a wide range of voltage ratings from 2.5 volts to 200 volts. For rating up to about 6 volts, they are available in increments of 0.1 volt each; from 6-17 volt ratings, in increments of 0.25 volts; from 17-30 volt ratings, in increments of 0.5 volts; and from 30 volts up, they are available in increments of 3-6 volt steps.

Most zener diodes are rated in watts, common values being 0.125, 0.150, 0.250, 0.400, 0.500, 0.750, 1, 1.5, 3, 5, 10 and 50 watts.

The zener diode conducts at a definite voltage. If it starts conducting at 25 volts, it may stop conducting at 24.9 volts for a particular load, and do so each time exactly and indefinitely.

The current required to operate the zener diode over its regulating range is very small. It may be as low as ½ of 1% of the zener current rating. However, for the best operating level, at least 10% of the zener current rating should be used. The current through the zener diode determines its exact voltage level. The zener diode is also affected by the temperature.

The zener diode is far superior to the gaseous regulator tube which it replaces in voltage regulator circuits.

One or more zener diodes connected in series will give a total zener voltage equal to the sum of the series string. For example, connecting a 10-volt zener diode in series with a 12-volt zener diode will give a zener voltage of 22 volts. Polarity should be observed with the arrows pointing in the same direction. Also, they should have approximately the same current rating.

Zener diodes themselves very seldom fail, so be sure to find the cause of failure by checking all other associated components, such as resistors, condensers, and transistors before replacing a zener diode.

Zener Diode Characteristics

Figure 5-2a shows a typical characteristic curve of many types of silicon diodes, where I represents the current, and E represents the voltage. The upper right curve is the normal diode connection with the positive to the anode and the negative to the cathode. The lower left curve is the

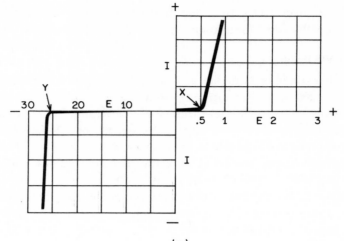

(a)

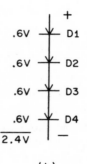

(b)

Figure 5–2   Zener diode characteristics

reverse, or zener, connection with the negative to the anode and positive
to the cathode. When connected in the forward direction, that is, with the
positive to the anode and negative to the cathode, a diode will not conduct
until the voltage has reached the barrier, or junction, voltage at point X
on the curve, and then the current will rise rapidly. This is the normal
diode operating range. However, when connected in reverse polarity, the

diode does not conduct until it reaches a much higher voltage, called the avalanche, zener or maximum PIV at point Y on the curve. Here the current rises even more rapidly than in the forward direction. Normally this region should be 2.5 to 3 times higher than the normal operating voltage of a normal diode. However, the zener diode is special in that it uses this characteristic as a very accurate voltage regulator control reference. A standard diode is sometimes used for this purpose when the voltage regulation is not critical. Where the zener diode is connected for reverse polarity, the normal diode is connected for forward polarity when used as a voltage regulator. Since one diode will conduct at the barrier, or junction, voltage at point X (about 0.6 volts), it will hold a fairly constant voltage over a wide range of currents on the straight part of the curve. Sometimes it is more economical to use several diodes in series instead of one costly zener diode to obtain voltage control at a level that does not require the accuracy of the zener diode. For example, Figure 5-2b shows four diodes D1, D2, D3, and D4 connected in series making a 2.4 volt voltage regulator capable of carrying much higher currents than available zener diodes.

Testing Zener Diodes

A zener diode can be tested for forward conductivity, opens, and shorts by using an ohmmeter. Small zener diodes, under 1 watt rating, will show resistance only on the RX10 ohmmeter scale, showing about 210K-250K ohms in one direction, and infinity (maximum resistance) in the opposite direction. The larger zener diodes, 1 watt and above, can be read on the RX1 scale, showing a very high resistance in the reverse direction and a very low resistance, less than 100 ohms, in the forward direction. This test is usually sufficient for service work.

The polarity test set shown in Figure 1-2 (Chapter 1, Section II) can be used, using prods PR1 and PR2, to test for polarity, opens, and shorts in the forward direction. Neither of these tests will give the zener voltage.

To determine the zener voltage, an adjustable D.C. voltage supply with a voltage range higher than the zener voltage is necessary. This can be supplied by test sets shown in Figures 1-7 and 1-8 (Chapter 1, Section II) or a separate test set can be built as shown in Figure 5-3. In this set, the battery BAT can be any steady, or filtered, D.C. source having a voltage higher than the zener voltage; F is a $\frac{1}{4}$ ampere fuse;

R1 and R2 are 1K ohm, 10 watt potentiometers; C is a 10 mfd, 100-volt electrolytic capacitor; V is a high resistance voltmeter; M.A. is a milliammeter, or the correct range of a V.O.M.; Z is the zener diode under test. Potentiometer R1 is wired to increase the voltage when turned clockwise, and R2 is wired to decrease resistance and increase the current, when turned clockwise.

The procedure to determine the zener voltage is as follows: with R1 and R2 turned to the extreme CCW position, connect the voltmeter V

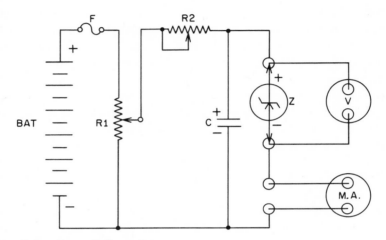

Figure 5–3   Zener diode testing

and the zener diode Z to the same terminals as shown. Determine the correct test current, select correct milliammeter range, and connect as shown. There should be no reading on the voltmeter V or on the milliammeter M.A. Slowly turn R1 in a CW direction and note the meters. The voltmeter should begin to show voltage but the milliammeter should show no current reading yet. If it does, the zener is either shorted or reversed and is conducting in the forward direction. Do not continue the test until this is corrected. Continue to turn R1 and watch the milliammeter M.A. As soon as it begins to read, continue to turn R1 CW until the zener diode "test current" value is reached. Read the voltmeter, which shows the zener voltage of the zener diode. If unable to reach the test current value after turning R1 fully CW, then turn R2 slowly CW until the test current is reached. Read the voltmeter V for the zener voltage. Rotate R1 or R2 slightly back and forth to increase and decrease the M.A. reading, and see that the voltage remains practically constant. The "test current" can be obtained from electronic supply catalogs if the type number is known or it can be calculated if the wattage rating is known. In Ohm's law, watts equal volts × amperes, or amperes equal watts/volts. If the wattage and zener voltage are known, then the maximum current can be calculated. The "test current" is usually given as about 25-40% of the maximum value, but may vary between different manufacturers. The "test current" is usually the same as the normal operating current for the specified zener voltage. For example, a 1 watt, 4.6 volt zener diode would carry a maximum current of 1/4.6 or .217 amperes, or 217 M.A.. The "test current" then would be 25% of 217 or 54 M.A.

SPECIAL SOLID STATE DEVICES AND TUBES

6

Transient Voltage Suppressors

Battery charger components, such as rectifiers, diodes, zener diodes, transistors, SCRs and other electronic parts usually operate within such close limits for normal operation that they cannot endure extremely high voltage peaks or "spikes" without damage. These spikes are called spurious or transient voltage surges. Transients may come into the power line from other sources, or be generated within the charger during switching.

Condensers, or resistor-condenser networks, are used effectively in suppressing these transient voltages, and are commonly used across the primary of a transformer, across the secondary, or across the rectifiers or rectifier output.

Small diodes are connected in reverse polarity across the solenoid of a battery charger alternator protector to discharge it when it is disconnected. Thus, transient voltages cannot damage the transistor or other solid state components.

The thyrector is a selenium rectifier used for intermittent operation in the reverse direction, that can protect other semiconductors from high voltage transients. The reverse characteristics are not as good as those of the silicon zener diode. So it is not a very good regulator, but it is sometimes used as such because it is inexpensive. It is known as a selenium surge protector, also. Since transients can occur on the A.C. lines, the D.C. lines, or both, selenium surge protectors are available for A.C. (non-polarized) as shown in symbol form in Figure 6-1a, and for D.C. (polarized) as shown in symbol form in Figure 6-1b. Of course, two

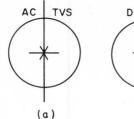

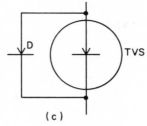

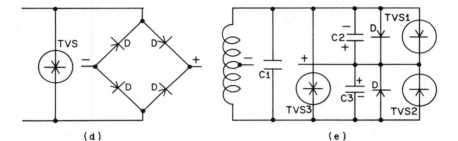

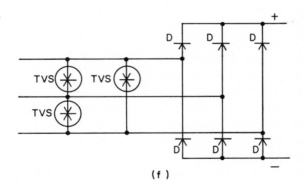

Figure 6–1   Transient voltage suppressors

polarized units as shown in Figure 6-1b can be connected back-to-back to equal the non-polarized unit shown in Figure 6-1a.

Varistors are voltage sensitive resistors, providing a decrease in resistance with an increase in voltage. But they require higher PIV semiconductors and higher steady state power consumption.

Zener, or breakdown, diodes are compact solid state devices that offer lower transient energy capacity and have a slower response to steep wave front transients than other more suitable devices.

Figure 6-1c shows a polarized unit TVS across the diode D in a half-wave rectifier circuit.

Figure 6-1d shows a non-polarized unit TVS on a full-wave bridge circuit.

Figure 6-1e shows two polarized units TVS1 and TVS2 in a center-tapped full-wave circuit, and the equivalent non-polarized unit is shown as TVS3. Sometimes a condenser C1 is added as shown or electrolytic capacitors C2 and C3 are placed in parallel with the diodes D with polarities as shown.

Any one, or a combination of the devices shown in Figure 6-1e may be used.

Figure 6-1f illustrates three non-polarized TVS units across each of the three phases in a 3-phase full-wave circuit.

Unless a charger was originally equipped with transient voltage devices, it is usually not necessary to add them except under unusual circumstances. Many modern rectifiers have built-in transient voltage protection. It is called "avalanche characteristics", and they act similar to a zener diode in that they can conduct current in the reverse direction over a limited range without damage. Those rectifiers without this feature draw a very high rate of current increase when the voltage reaches the zener point. This characteristic also permits longer and higher overloads in the forward direction.

Many times the transient voltage protection is provided, even with rectifiers having good avalanche characteristics, as an added precaution or for unusual applications.

Unijunction Transistors (UJT)

The unijunction transistor (UJT), which is a single junction transistor, is not commonly found in battery charger circuits, but it has possibilities in this field, and a knowledge of these devices may be valuable in the near future. They have different characteristics from those of the conventional NPN and PNP transistors, which are two-junction devices. They feature a stable triggering voltage VP which is a fixed fraction of the voltage applied between the two bases, a very low value of firing current IP, a negative resistance characteristic, a high pulse

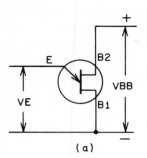

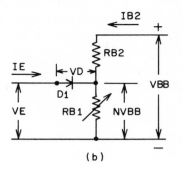

(a)

(b)

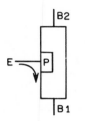

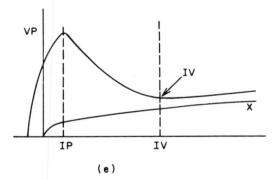

(c)

(d)

(e)

Figure 6-2 Unijunction transistor (UJT)

current capability, and low cost. These characteristics make them useful in voltage regulators, oscillators, timing circuits, voltage and current sensing circuits, SCR trigger circuits, and stabilizing circuits.

The symbol of the unijunction transistor is shown in Figure 6-2a, and the electrical equivalent circuit in Figure 6-2b. RB1 and RB2 represent the total resistance RBB between base 1 and base 2 (between 5K and 10K ohms). The silicon bar shown in Figure 6-2c acts as a resistance voltage divider between base B1 and base B2. A single rectifying contact, called emitter E, is made between base B1 and base B2. The diode D1 in Figure 6-2b represents the the unijunction transistor emitter diode. Usually RB1 is about 20% higher than RB2. RB1 is shown as a variable because its resistance varies with the emitter current. In normal circuit operation, base B1 is grounded, and a bias voltage VBB is applied with positive polarity at base B2 and negative polarity at base B1. With no emitter current flowing, the silicon bar acts as a simple voltage divider and a certain fraction N of VBB will appear at the emitter. If the emitter voltage is less than NVBB, the emitter will be reverse biased and only a small leakage current will flow. However, if VE becomes greater than NVBB, then the emitter will be forward biased and emitter current will flow from the emitter E to base B1. This results in a decrease in the resistance between the emitter E and base B1. So, as the emitter current increases, the emitter voltage decreases. This is negative resistance. Negative resistance is obtained when a voltage decrease results in an increase in current, or an increase in current results in a voltage decrease. This is just the opposite to Ohm's law in normal circuits.

The characteristic curve in Figure 6-2e shows the peak voltage VP and peak current IP when the emitter E starts to conduct, and the valley point of current IV. The region to the left of the peak emitter current point IP is called the "cut-off" region; here the emitter is reverse biased and only a small leakage current flows. The region between the peak point IP and the valley point IV is the negative resistance region. The region to the right of the the valley point IV is the saturation region; here the dynamic resistance is positive. The lower curve X shows only the emitter-to-base B1 diode characteristics, with no base current through base B2.

Figure 6-2d shows the electrical connections to the terminals of the unit.

Silicon Controlled Switch (SCS)

The silicon controlled switch (SCS) is a new member of the transistor family. Therefore, it has not been commonly used in battery charger

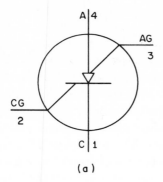

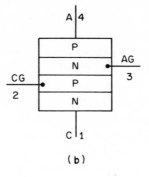

(a)

(b)

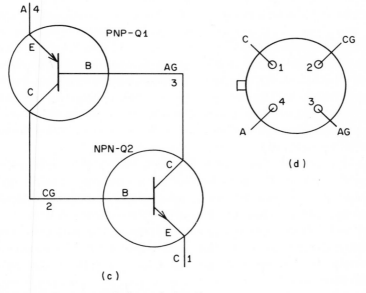

(c)

(d)

Figure 6–3 Silicon controlled switch (SCS)

control circuits, but it has possibilities in this field of electronics. A working knowledge of this device will be helpful in understanding any future circuits in which they are used.

The SCS is a PNPN structure with all four semiconductor regions accessible as shown in Figure 6-3b, rather than only three as is customary with the SCR. This greatly expands circuit possibilities beyond those of conventional transistors and SCRs.

The SCS symbol (Figure 6-3a), structure (Figure 6-3b), and equivalent circuit (Figure 6-3c) are the same as those of the SCR except for the addition of an anode gate AG-3. The SCS operates like an SCR except that it can also be triggered into conduction by applying a negative bias on the anode gate AG-3 between anode A-4 and cathode C-1. It also has several specialized modes of operation.

The nearest electrical equivalent of the SCS uses one PNP-Q1 transistor and one NPN-Q2 transistor connected as shown in Figure 6-3c. This circuit is often used in some voltage regulator control circuits of battery chargers. A single SCS could therefore possibly replace the two transistors by proper circuit design.

Figure 6-3d shows the base connections as seen from the lead side.

Silicon Unilateral Switch (SUS)

The silicon unilateral switch (SUS) is essentially a minature SCR, having an anode gate like the SCS (instead of the usual cathode gate) and a built-in low voltage avalanche diode (zener Z) between the gate G and the cathode C. With a negative bias on the gate G, the SUS is triggered into conduction at the zener section voltage level (8 volts, for example).

Figure 6-4a shows the symbol of the SUS; Figure 6-4b shows the structure of the SUS; and Figure 6-4c shows the nearest equivalent circuit of an SUS, where the zener diode Z is connected across the anode gate diode D, much like the SCS less the cathode gate.

The SUS has many future uses in voltage regulator circuits, but until now has not found common usage in battery charger circuits. However, a working knowledge will be helpful if, and when, it is used in battery charger circuits.

Figure 6-4d shows the base connections seen from the lead end.

Gate Controlled A.C. Switch (TRIAC)

A gate controlled A.C. switch (triac) shown in Figure 6-5 operates like an SCR except that it can be triggered into conduction in either direction by either a positive or a negative gate G signal.

Figure 6-5a shows the triac symbol, Figure 6-5b shows the PNPN structure, and Figure 6-5c shows the electrical equivalent of the triac, which

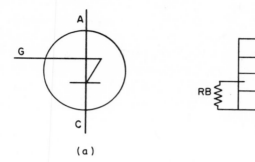

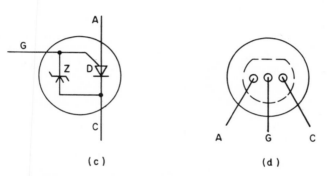

Figure 6–4 Silicon unilateral switch (SUS)

is the same as two SCRs connected in inverse parallel. Figure 6-5d shows the terminal connections to the triac. The type of mounting may be either by stud, as shown, or by press fit.

Because the triac can conduct current in both directions, it cannot be turned off by reversing the current. If the voltage across a triac were instantaneously reversed, the recovery current that flowed would merely turn the device on in the opposite direction. To turn a triac off successfully, the current through it must be reduced below the holding current by reducing the applied voltage to zero. Then, sufficient time must pass before the reapplication of voltage, in either direction, to allow natural recombination of any stored energy charge. In 60 Hz A.C. circuits, which is the main use of the triac, the rate of change of voltage near zero current of each cycle allows proper operation in this manner.

At the present time, triacs are not commonly used in battery charger

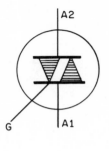

(a)

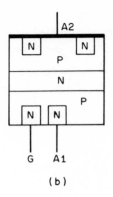

(b)

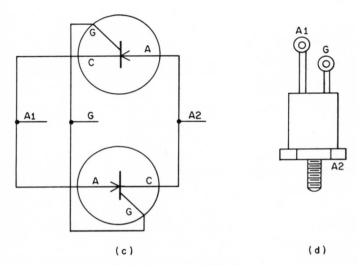

(c)

(d)

Figure 6–5   Gate controlled AC switch (TRIAC)

circuits, but they have possibilities in this field, and a knowledge of these units will help when they are commonly used. For example, the triac can act as an A.C. switch actuated by the battery voltage regulator, and eliminate relays.

Silicon Bilateral Switch (SBS)

The silicon bilateral switch (SBS) is essentially two silicon unilateral switches (SUS) arranged in inverse parallel. Since the SBS operates

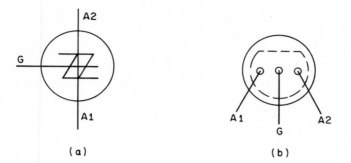

(a)   (b)

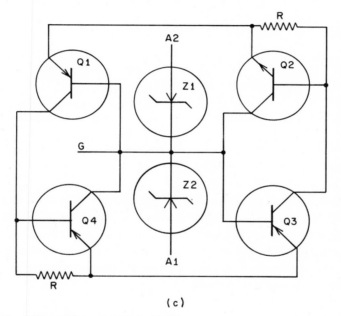

(c)

Figure 6-6   Silicon bilateral switch (SBS)

as a switch with both polarities of applied voltage, it is especially useful
for triggering the bidirectional triode thyristors (triacs) with alternate
positive and negative gate pulses. It is ideally suited for half-wave and
full-wave triggering in low voltage SCR and triac phase control circuits.

Figure 6-6c shows the electrical equivalent of the SBS using four
transistors Q1, Q2, Q3, and Q4, and two zener diodes Z1 and Z2. The
two built-in zener diodes give an accurate voltage reference point, and

are stable over normal temperature ranges. A gate lead gives access to the zener and PNP base node as shown.

Figure 6-6a shows the SBS symbol, and Figure 6-6b shows the SBS electrical connections.

Although the SBS is not commonly used in battery charger circuits now, it will be helpful to know the symbol and connections should they be encountered.

Diode A.C. Switch (DIAC)

The diac is both a bilateral trigger diode and a diffused silicon bi-directional trigger diode, which may be used to trigger the triac or SCRs.

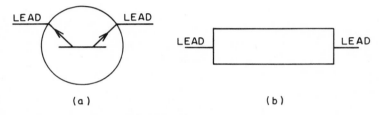

LEAD   LEAD   LEAD   LEAD

(a)   (b)

Figure 6–7   Diode AC switch (DIAC)

This device has a three layer structure similar to a transistor and has negative resistance characteristics above a certain switching current for both directions of applied voltage.

The diac symbol is shown in Figure 6-7a, and the electrical connections are shown in Figure 6-7b. The unit is bi-lateral and can be connected either way, since it has no polarity.

Thyratron Tubes

The thyratron tube is a gas-filled glass tube with a heater or filament, a grid and a plate, and sometimes a screen grid. They were used in early battery chargers as a voltage regulator, but are now being replaced by solid state devices. There are still many thyratrons in service and replacement parts are available. Also, solid state thyratrons may replace the tube type. A thyratron tube acts like an SCR in that once it is fired to conduct, the control grid has no control over the anode-to-cathode current. In the SCR the gate has no control over the anode-to-cathode current once it is fired.

The two most common types of thyratrons used in battery charger control circuits are the 2D21 and 2050. They have a heater voltage of 6.3 at a current of 0.6 amperes, and are the shielded grid type. Each has an anode current of 100 M.A. average. The 2D21 will handle a peak anode

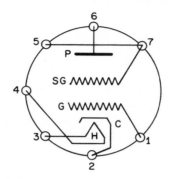

Figure   6–8   Thyratron tube base connections

current of 500 M.A. and the 2050 is rated at a peak anode current of 1 ampere.

Figure 6-8 shows the thyratron tube base connections, where P is the plate or anode, SG is the screen grid, G is the control grid, C is the cathode, and H is the heater.

Although the thyratron is obsolete for battery charger voltage regulators, they are still used in oscilloscope-type engine analyzers, and other related applications.

CAPACITORS AND CONDENSERS

7

Capacitors are used to store an electrical charge. Also, they are used to reduce the undesirable effects of transient voltage spikes, to provide voltage feedback circuit elements, to filter out alternating current or pulsating direct current, and to correct or adjust the power factor in a power line. Capacitors are made in many different sizes and shapes. Depending on the materials used to make them, capacitors are sometimes called condensers. Generally, capacitors use an electrolytic (chemical) dielectric or insulator while condensers use a paper or oil-type dielectric.

All capacitors or condensers are electrically and schematically alike (Figure 7-1). A capacitor is made of two metal plates or sheets of aluminum foil separated by an insulator. The insulator is any non-conductive material, even air. The most common insulators are chemically treated paper, mica, mylar (plastic film), and oil.

Electrolytic capacitors are made of two sheets of aluminum foil, separated by a chemically treated insulation or dielectric. This chemical "sandwich" is rolled up into a cylinder and a terminal wire attached to each plate. The capacitor is polarized so that one terminal or plate is positive and the other plate or terminal is negative. The positive plate has more free electrons than the negative plate. This type capacitor must be installed with proper regard for polarities, positive to positive D.C. A special quality of this capacitor is that the plate-to-plate resistance is very high in one direction, and very low in the other direction. It is used only in direct current circuits. If two polarized capacitors are connected back-to-back (positive leads connected together and negative leads used as the capacitor leads), a non-polarized capacitor is formed that may be used in alternating current circuits for a short time or at reduced voltages for longer times. A nonpolarized capacitor built into one unit is often used as an electric motor starting capacitor.

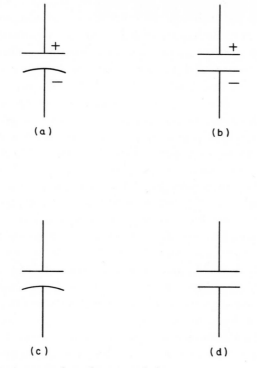

Figure 7–1 Capacitor and condenser symbols

Oil or paper condensers are classified or rated by size (capacity), working voltage, and the maximum or peak voltage that the capacitor can take before the dielectric breaks down. The size or capacity of the capacitor or condenser is rated in units called farads or some part of a farad. The most common size is in the microfarad range or the micro microfarad range. A microfarad (mfd) is one-thousandth of a farad; and a micromicrofarad (mmfd) is one-millionth of a farad. The capacity and the working voltage (either dc or ac) are usually marked on the outside of the capacitor or condenser. Early marking systems used a color coded dot system to identify the size and voltage limits. The color code was similar to the resistor color code in number values.

Since capacitors store up an electrical charge or current, circuit current is blocked from passing through the capacitor. However, the effect of the voltage is as if it passed through the capacitor. Actually, the voltage passes around the dielectric through the other circuit components to the other side of the capacitor. Voltage surges or spikes are absorbed and gradually discharged through the circuit resistive components. This charging and discharging of the capacitor takes time and the voltage is slightly delayed behind the current. Since the current leads the voltage on a capacitor, the capacitor has a leading power factor. This power factor is expressed as a percentage of the power or watts used by a circuit. The power in a circuit is a product of the voltage multiplied by the instantaneous "in-phase" current. The circuit volt-amperes is the product of the voltage times the amperes. The power factor is the percentage obtained by dividing the circuit power by the volt-amperes. The difference betweeen the power factor amperage and the full power amperage is "wattless" current that does not register on a kilowatt hour meter. The power company does not get paid for providing transformer and transmission line capacity to carry this extra wattless current, and therefore offers special rates to industrial plants to correct the power factor. This is usually done by static oil—filled condensers, or by using large synchronous motors which operate with a strong or over excited field that acts as a condenser across the line.

Large industrial battery chargers, or rectifier units, may use a condenser connected across the line or a high voltage secondary to correct the power factor. The power factor is not corrected on smaller 115 volt chargers. On these units, a small condenser may be connected across the primary winding to suppress high transient voltage spikes in the line.

Condensers are frequency conscious; that is, the higher the A.C. frequency the more conductive they become. Transient voltages may be high voltage, and high frequency. A steep fronted spike or wave shape contains high frequencies, and is easily absorbed by the condenser. Condensers store the high voltage peaks and discharge them back into the circuit as the voltage reduces. Thus, the rectified current battery charger is smoothed out, and the filter action absorbs transient voltages. Capacitors are used to smooth out, or steady, rectified voltages to stabilize operation, and prevent spurious triggering of voltage regulator circuits using zener diodes, SCRs and transistors, that might be damaged by transient voltages.

Capacitor or condenser symbols used in schematic wiring diagrams are shown in Figure 7-1. Figure 7-1a shows the symbol for shielded type electrolytic capacitors with the mfd value, D.C. working voltage (DCWV), and the polarity. Figure 7-1b shows the symbol for unshielded electrolytic capacitors. Figure 7-1c shows shielded type paper, mica, or oil-filled condensers. The mfd and VAC are indicated with the symbol on schematics. Figure 7-1d shows an unshielded type of condenser.

The shielded type condenser is important in battery charger circuits only when the equipment is in the vicinity of a powerful radio transmitter. In such circuits employing thyratrons or SCRs, radio signals can trigger the circuit into unwanted operation. Some thyratron circuits even use a radio frequency choke to block radio signals.

A capacitor and condenser checker is necessary in battery charger service. They are available in factory built, or in kit form from any electronic supply house. They determine mfd value, leakage and power factor of electrolytic capacitors, and mfd value and leakage of oil or paper condensers.

Any capacitor or condenser has some voltage leakage caused by internal parallel and series resistance. Voltage leakage is extremely low (high-resistance) in paper, mica or oil-filled condensers, but electrolytic capacitors have more leakage and are tested for a power factor rating. A high power factor reading indicates a high power loss and low efficiency. A pure resistance, and no capacitance, would have a unity power factor of 1.0. A perfect condenser, or capacitor, would have zero power factor, and all the amperage would be wattless. Therefore, the lower the power factor the more efficient the unit would be. Lacking more specific tolerances, an electrolytic capacitor with more than 25% power factor should be replaced depending on its place in the circuit. A high power factor reduces filtering action and causes hum in a filter circuit even if the mfd value is correct.

Identical Condensers in Series, or Parallel

When identical condensers are connected in series, the voltage rating of the combination is the sum of all the voltages. Multiply the voltage of one unit by the number of condensers in series to get the total voltage of the combination. Divide the mfd value of one unit by the number of units in series to get the total mfd value of the combination. For example, in Figure 7-2a, the two condensers C1 are 1 mfd each at 330 VAC. The value of C is 0.5 mfd at 660 VAC.

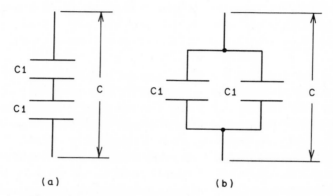

(a)   (b)

Figure 7–2   Identical condensers connected in series or parallel

When identical condensers are connected in parallel, the voltage rating of the combination is the same as for a single unit. The mfd value is the mfd value of one unit multiplied by the number of units. For example, if Figure 7-2b C1 is 1 mfd 330 VAC, the total value of C would be 2 MFD at 330 VAC.

Dissimilar Condensers in Series, or Parallel

Figure 7-3a shows a series arrangement of two or more condensers of different mfd values. Use the following formula to calculate the total mfd value of C.

$$C = \frac{1}{\dfrac{1}{C1} + \dfrac{1}{C2} + \dfrac{1}{C3}}$$

For only two condensers in series use the following formula.

$$C = \frac{C1 \times C2}{C1 + C2}$$

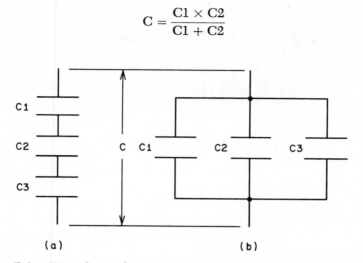

Figure 7–3 Dissimilar condensers connected in series or parallel

Figure 7-3b shows two or more condensers of different mfd value connected in parallel. To get the total mfd value C, add C1, C2 and C3. The voltage rating will be the same as a single condenser.

RESISTORS AND THERMISTORS

8

Shapes and Symbols

Resistors used in battery charger circuits have many shapes and ratings as shown in Figure 8-1.

Figure 8-1a shows the symbol of a fixed value resistor, as it appears in schematic wiring diagrams. The symbol does not show the type of resistor, but usually the resistance in ohms and the wattage rating are marked next to the symbol.

Figure 8-1b and Figure 8-1c show typical small carbon composition resistors, usually limited to wattage ratings of ¼, ½, and 1 watt, and made in any resistance value. They are used in alternator protectors, meter multipliers, and voltage regulator circuits. Because of physical size they usually do not have the ohm value marked on the resistor, but use a RMA color code system. The earlier color code system, shown in Figure 8-1b, indicates the resistance as follows:

> Body color (A) = 1st digit
> End color band (B) = 2nd digit
> Middle color band or dot (C) = number of zeros to follow the
> first two digits
> Tolerance band (D = Gold—5%; Silver—10%;
> no color band—20%.

The present color code system widely used for carbon resistors is as follows:

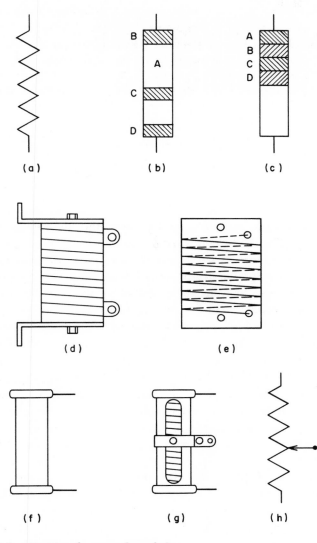

Figure 8–1 Resistor shapes and symbols

Color band (A) = 1st digit
Color band (B) = 2nd digit
Color band (C) = number of zeros to follow the first
two digits
Color band (D) = tolerance band—Gold, 5%;
Silver—10%; no color band—20%.

The various colors used in both coding systems represent the numbers given below:

Black-0	Green-5
Brown-1	Blue-6
Red-2	Violet-7
Orange-3	Gray-8
Yellow-4	White-9

The tolerance band on each resistor gives the acceptable limits of measured resistance for that resistor. For example, a resistor with a color code of 100 ohms with a 20% tolerance may measure from 80 to 120 ohms and still be acceptable. If this same 100-ohm resistor had a tolerance of 10%, it could measure from 90 to 110 ohms and still be acceptable. And if this same resistor had a tolerance of 5%, it could measure from 95 to 105 ohms and still be acceptable. However, if any resistor measures a higher or lower resistance than the indicated tolerance allows, discard the resistor and install a new one.

A typical wire wound resistor is shown in Figure 8-1d. This resistor is made of a length of resistance wire that is wound around a cylindrical porcelain spool. It is used in battery charger circuits that allow only a small amount of current to trickle to the battery being charged.

Another type of wire wound resistor is shown in Figure 8-1e. It is basically the same as the resistor shown in Figure 8-1d but the wire is wound on a flat porcelain or asbestos plate.

Both types of wire wound resistors often are not marked with the resistance or wattage values but they may have a manufacturer's part number marked on them. Wire wound resistors may be identified by measuring their resistances or by tracing the circuit wiring of the battery charger unit and comparing it with a wiring diagram. If a wiring diagram of the battery charger unit is not available and a replacement resistor is needed, the proper resistance may be estimated by measuring any part of the wire winding that is undamaged using an ohmmeter. The measured resistance is then multiplied by the number of times required to make a whole resistor. For example, if only one-fourth of the total winding of a resistor can be measured, multiply the resistance of that portion of the winding by four to get the total approximate resistance of the whole resistor. These wire wound resistors often range from 6 to 20 ohms with wattage ratings of 50 to 100 watts.

Figure 8-1f shows a tubular, or square, wire wound resistor that usually has the ohms and watts rating printed on it.

Figure 8-1h shows the symbol for an adjustable wire wound resistor.

The tubular resistor with its exposed wire winding (Figure 8-1g) is an adjustable resistor. The sliding contact is set and clamped in place when the desired resistance is read on an ohmmeter connected between the end terminals and the contact arm. Sometimes the sliding contact and the wire winding need to be cleaned with a fine sandpaper when an open (no continuity) condition is present in the resistor. If the cleaning operation does not correct the problem, moving the sliding contact to another part of the resistor may allow further use of the resistor.

Sometimes it is necessary to calculate the resistance, current, voltage, and wattage conditions in a circuit. If any two values are known, the third unknown value can be found using the Ohm's law formulas given below.

1. Unknown voltage (E) = current (I) × resistance (R)
$$\text{or } E = I\,R$$

2. Unknown current (I) = voltage (E) ÷ resistance (R)
$$\text{or } I = \frac{E}{R}$$

3. Unknown resistance (R) = voltage (E) ÷ current (I)
$$\text{or } R = \frac{E}{I}$$

4. Unknown watts (W) = current (I) × voltage (E)
$$\text{or } W = I\,E$$

5. Unknown watts (W) = current (I) × current (I) × resistance (R)
$$\text{or } W = I^2\,R$$

6. Unknown current (I) = square root of [watts (W) ÷ resistance (R)]

$$\text{or } I = \sqrt{\frac{W}{R}}$$

For example, a 10-ohm, 100-watt resistor has a 31.63 voltage drop with a 3.163 ampere current. Substituting these values in the above formulas, they can be solved as follows:

$$W = EI = 31.63 \times 3.163 = 100 \text{ Watts}$$
$$E = IR = 3.163 \times 10 = 31.63 \text{ Volts}$$
$$I = \frac{E}{R} = 31.63 \div 10 = 3.163 \text{ Amperes}$$
$$R = \frac{E}{I} = 31.63 \div 3.163 = 10 \text{ Ohms}$$
$$W = I^2R = 3.163 \times 3.163 \times 10 = 100 \text{ Watts}$$

$$I = \sqrt{\frac{W}{R}} = \sqrt{\frac{100}{10}} = \sqrt{10} = 3.163 \text{ Amperes}$$

Find the square root of a number by using a slide rule, a mathematics book, or Machinery Handbook, or find it as follows: to take the square root of a number under the square root sign $\sqrt{}$, find a number that when multiplied by itself will equal the number under the square root sign. For example, what is the square root of 120 $\sqrt{120}$? This can be figured in several operations by trial and error, and get close enough for all practical purposes. Try 10×10 is 100, 11×11 is 121, so the number is very close to 11, and is close enough for most purposes. The actual square root of 120 is 10.9545.

Resistor Circuits, Series and Parallel

Resistors are connected in circuits in series and parallel arrangements. Series resistor circuits have all the resistors connected end-to-end as shown in Figure 8-2a. All the current passes through each resistor and the voltage is divided among all the resistors according to size. Large resistors use more voltage than small resistors. Large resistors also use more wattage than small resistors. The following statements apply to resistor circuits:

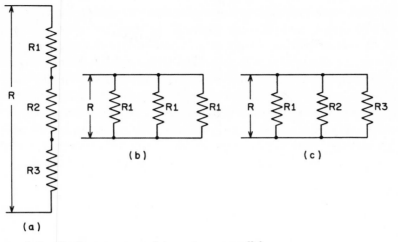

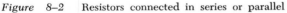

Figure   8–2   Resistors connected in series or parallel

SERIES RESISTOR CIRCUITS.

1. Total resistance is the sum of all resistors in the circuit, or

$$Rt = R1 + R2 + R3.$$

2. Total voltage drop is the sum of all voltage drops across the resistors, or

$$Et = E1 + E2 + E3.$$

3. Total wattage (W) is the sum of all the watts dissipated by each resistor, or

$$Wt = W1 + W2 + W3.$$

4. If all the resistors are the same size, the values (resistance, voltage, current, etc.) may be multiplied by the number of resistors in the circuit.

5. If voltage and resistance are known, use formula $W = \dfrac{E^2}{R}$ to find the wattage.

6. If current and resistance are known, use formula $W = I^2 R$ to find the wattage.

Parallel-connected resistors are connected as shown in Figures 8-2b and 8-2c. When two or more identical resistors are connected in parallel, the total resistance is equal to a part of one resistor. For example, if there are two resistors in the circuit, the total resistance equals $\frac{1}{2}$ of one resistor. If there are three resistors in the circuit, the total resistance equals $\frac{1}{3}$ of one resistor. And if there are four resistors in the circuit, the total resistance equals $\frac{1}{4}$ of one resistor. The total wattage of such a circuit is the sum of all the wattages of the resistors.

To calculate the total resistance of a parallel-connected circuit of two or more different sized resistors, use the following formulas:

1. Two different sized resistors

$$Rt = \frac{R1 \times R2}{R1 + R2}$$

2. Three or more different sized resistors

$$Rt = \frac{1}{R1 + R2 + R3}$$

The wattage of each resistor should be calculated to avoid overloading it. If the voltage and resistance are known, use the formula $W = \frac{E^2}{R}$ to find the wattage. Add all the individual wattages to get the total wattage of the circuit. Remember that the voltage on each branch of a parallel circuit is the same as the others.

Thermistors

The thermistor is a special type of resistor that changes its resistance as the surrounding temperature changes. However, the thermistor has a negative temperature coefficient that causes its resistance to vary opposite to the temperature change. That is, as the temperature rises, the thermistor resistance decreases; and, as the temperature decreases, the thermistor resistance rises. By proper selection of resistors and thermistors, the circuit resistance can be held constant over a wide range of temperature changes. If it were not for thermistors, equipment, such as a TV set, would need constant adjustment as it warmed up. In battery charger voltage regulator circuits, the thermistor is used to compensate for temperature changes and to over-compensate for temperature changes to provide a lower cut-off voltage at higher temperatures and a higher cut-off voltage at lower temperatures. Figure 8-3 shows the 3 most common thermistor schematic symbols.

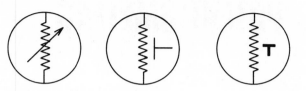

Figure 8–3   Common thermistor symbols

9
RHEOSTATS AND POTENTIOMETERS

Rheostats and potentiometers are different forms of variable resistors. Each has a movable contact that wipes over the surface of a resistance element, which may be a carbon compound or wire wound. They can be adjusted from zero resistance to the maximum resistance value. All that was said under the subject of resistors in Chapter 8 applies to rheostats and potentiometers. The only difference between a rheostat and a potentiometer is the number of terminals and how it is connected in the circuit.

Figure 9-1a shows the rheostat symbol as used in schematic wiring diagrams. Figure 9-1b shows the general appearance, from the back side, of a wire wound rheostat.

Figure 9-2a shows the potentiometer symbol, and Figure 9-2b shows the general appearance of a typical wire wound potentiometer.

Figure 9-2c and d shows how a potentiometer can be used as a rheostat by using only terminals 1 and 2, or 2 and 3. However, if terminals 1 and 2 are being used, always connect a jumper wire J between terminals 2 and 3. Jumper wire J will carry part of the load and prevent a completely open circuit if the wiper is lifted from the resistance element momentarily by a piece of foreign matter or dirt. Jumper wire J is necessary to prevent constant arcing and to prolong the life of the rheostat. The jumper wire is used only when a potentiometer is used as a rheostat.

A rheostat has one connection to the resistance element, and one to the wiper, or movable contact. It is identified as a rheostat in the parts list but may be supplied as a potentiometer.

Rheostats and potentiometers used in battery charger circuits are usually "linear", that is, the resistance increases at a constant rate over the entire range, having the same ohms per degree of shaft turn. However, the non-linear, resistance element, having more resistance at one end than the other, is used sometimes.

Rheostats and potentiometers are made with a variety of adjusting shafts depending on their intended use. Frequently-adjusted rheostats

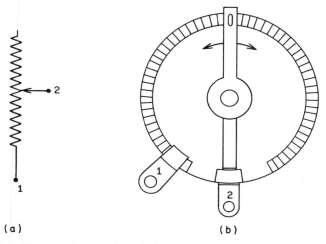

(a)   (b)

Figure 9–1   Rheostat shape and symbol

and potentiometers have shafts fitted with a knob and are accessible from the front of the battery charger. Rheostats and potentiometers used for meter calibration and voltage regulation adjustments have shafts that are slotted in the end and a locknut to hold the setting steady. This type of control is usually located inside the battery charger and may be accessible through a hole in the case. This type of adjustment should be made by a properly trained technician only. Most rheostats and potentiometers are panel-mounted through a drilled hole just large enough for the threaded mounting shaft. The component is installed from the rear of the panel and the lock washer and locknut are put on from the front of the panel.

These units are easily tested for continuity, smoothness of operation,

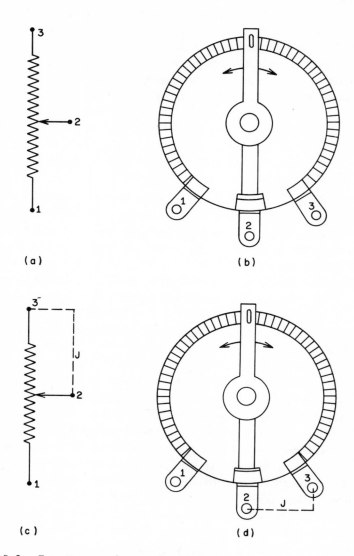

(a)

(b)

(c)

(d)

Figure 9–2 Potentiometer shape and symbol

and resistance value using an ohmmeter. The resistance, wattage, and current values are usually marked on the unit. The safe current carrying capacity and wattage are important to know. The wattage rating is determined by the total safe watts it can dissipate with all of the resistance in the circuit. Likewise the maximum safe current must be limited to no greater current than flows through the whole resistance at full wattage.

For example, a ½ ohm 50 watt wire wound rheostat would have a maximum current carrying capacity of:

$$I = \sqrt{\frac{W}{R}} = \sqrt{\frac{50}{.5}} = 10 \text{ Amps.}$$

This must not be exceeded at any point on the resistance element. It will dissipate only 50 watts with all of the resistance in the circuit. At one-half scale it would have ¼ ohm resistance and could dissipate only 25 watts at 10 amperes. The resistance wire will carry only 10 amperes maximum.

Rheostats are usually connected in series with the transformer primary A.C. input, or D.C. rectifier output, to give a "fine" adjustment of the charging rate between the "coarse" settings provided by tap switches. Also they are used in series with fixed multiplier resistors to make calibration adjustments on the meter circuits.

Potentiometers are used in voltage regulator circuits to provide an adjustment of the cut-off, or turn-on, voltage of the charger, by feeding a predetermined voltage to a zener diode, thyratron, or other control device. Terminals 1 and 3 are connected across a voltage source, with movable contact 2 supplying an adjustable reduced voltage. For example, a 1000 ohm potentiometer connected across 12 volts at terminals 1 and 3, and a voltmeter connected to terminals 2 and 3, or 2 and 1, could be made to read between zero and 12 volts by turning the potentiometer shaft.

10
BATTERIES

Battery Types

There are many types of batteries designed for many purposes. Some batteries are classified as rechargeable. These are called secondary cells, or storage batteries. Other batteries are not considered rechargeable, but they can be restored, or rejuvenated, by charging for a limited number of times, and with various degrees of success. These are called primary cells.

The most common rechargeable battery types are the lead-acid storage battery, the nickel-iron alkaline (Edison type) storage battery, and the nickel-cadmium battery.

The most common types considered as not truly rechargeable are alkaline, mercury, silver-cadmium, carbon-zinc (ordinary dry cell) and silver oxide.

Lead-Acid Storage Batteries

The lead-acid storage battery is commonly used for short term, high current applications, such as vehicle starting, as well as for medium and low current applications over longer periods of time, such as for trolling motors, golf carts and so forth.

The main advantages of lead-acid batteries are low internal resistance and low cost. The main disadvantages are weight and short life.

A lead-acid battery is made up of a number of positive plates, an equal number of negative plates, and a number of insulating separators placed between the positive and negative plates. This sandwich, of each cell, is placed in a separate compartment of a case made of insulating material that will withstand the sulphuric acid solution filling each cell.

A cell can be made, and the construction understood, by immersing two lead plates in a solution of sulphuric acid and water, having a specific gravity of 1260. Voltage will not be developed until the battery is polarized by charging. The plate connected to the positive clamp of the battery charger will gradually have a deposit of lead peroxide formed on its surface, while the plate connected to the negative clamp of the charger will gradually build up a deposit of pure lead on its surface. When the battery is discharged, part of the sulphur dioxide in the electrolyte solution is absorbed by the plates, forming lead sulphate deposits and lowering the specific gravity of the electrolyte. If the battery is left in a discharged condition for a long time, this lead sulphate becomes permanent and cannot be changed back to lead and lead peroxide by charging. The battery is said to be "sulphated", and cannot take a full charge. As the battery is charged and discharged over a period of time, a small amount of the plate material loosens and drops to the bottom where it builds up, eventually shorting the cell, discharging it while the remaining cells remain charged. There are battery rejuvenator chemicals that may dissolve this deposit, but they are of questionable value, and are expensive. Most of them use plain Epsom salts as a base. Try one or two teaspoons of Epsom salts in each cell, and in some cases, the deposits will be gone in a few days.

Commercial battery plates are lead grids with a paste mixture of lead peroxide pressed tightly in the grid structure to form the positive plate, and a paste mixture of pure lead pressed into the negative grid. The separators are wood, fiberglass or hard rubber, and ribs to permit the free flow of electrolyte. A solution of sulphuric acid and water, having a specific gravity of 1260, must cover the plates at all times. Only the water evaporates, so it is never necessary to add acid, unless some is spilled. The battery is charged and discharged several times before putting it into full service. Previously, the electrolyte was put in at the factory, and had a limited life, even though it was not in use. Batteries held in stock required a small maintenance charge of less than one ampere to keep the battery from permanently setting up a sulphated condition. Recently, however, the batteries are shipped "dry charged", requiring the addition of the electrolyte at the time it is put into use. The battery needs only a short charge after adding the electrolyte. Dry charge instructions usually specify that, after adding the electrolyte and allowing a five minute soaking time, a charge of 50 amperes be applied for 10 minutes for a 6 volt battery, 30 amperes for 10 minutes for a 12 volt battery, or a charge of 6 amperes for one hour. Add enough electrolyte to bring the solution level $\frac{1}{8}$ inch above the plates.

Dry charge batteries do not deteriorate while dry, and can be stored for long periods of time without any attention.

Lead-acid storage batteries have a predictable useful life, and usually carry a guarantee. The limiting factor is primarily the material used in the separators and the construction of the plate grids. The useful life may be from 1 to 5 years, or more, depending on use and care. Vibration and lack of proper maintenance, such as, overcharging or undercharging are the greatest enemies of battery life.

For example, batteries used on golf carts are designed to give more ampere hours at a low to medium current drain of 25 to 75 amperes (75A. for 75 min. is a standard) where a vehicle starting battery is designed to give maximum ampere-hour capacity at higher currents of 200-400 amperes for only a few seconds or minutes at the most.

To charge a lead-acid battery properly, the battery should start charging at the 8 hour rate. If the ampere-hour capacity of the battery is known, divide the figure 8 (8 hrs.) into the ampere-hour rating. For example, a 100 ampere-hour battery would be charged at 100 divided by 8, or 12.5 amperes charging rate. The charge will taper off to a lower value unless the charger is of the "constant current" type. When the battery is fully charged, the electrolyte specific gravity should be 1260 as read on a battery hydrometer.

Fast charging is usually necessary to save time, but it is not good for the battery if done too often, or for too long a period of time. Fast charging should be limited in time, depending on the state of charge. Usually, charger timers mark the time in minutes for a given electrolyte reading and the battery voltage. Also, some fast rate chargers have a thermostat placed in one of the cells to shut off the charger when the electrolyte temperature reaches 125 degrees F. which is far below the boiling point. Other automatic chargers monitor the battery voltage and use a voltage regulator circuit and relay to shut off the charger when the battery voltage indicates a full charge.

Lead-acid batteries should be completely discharged occasionally, and then, after waiting ½ hour to normalize, completely recharged. Occasionally, put the battery on a long charge at a low rate to de-sulphate and equalize the cells.

A completely sulphated battery will act as an insulator, and will not show a charging current. This is important to know in series or multiple battery charging. The sulphated battery must be located and removed to charge the others. Some batteries may be so badly sulphated or discharged that they will not operate the solenoid on chargers equipped with an alternator protector. Some chargers have a "low battery actuator switch" which shunts the control to apply a low charge sufficient to operate the alternator protector when the switch is released.

The lead-acid battery has a voltage of 2.1 volts per cell when fully charged and in good condition. Each cell will maintain a voltage of approximately 2 volts at no load at practically any state of charge.

When a battery is under charge, the voltage builds up in proportion to the state of charge, reaching a leveling off point when the battery is fully charged and all cells are gassing freely. This leveling off voltage point is determined by the rate of charge, being higher for high rates of charge, and lower for low rates of charge. Also, the leveling off voltage is lower at high temperatures, and higher at low temperatures.

Automatic voltage regulators that operate to reduce or cut off the charge use this leveling off voltage as a reference point. For vehicle regulators, the voltage is held at a point just below the gassing point to avoid boiling off excessive amounts of water. For golf cart chargers, and the like, when this gassing voltage is reached, the charger is cut off completely, or the charge rate is reduced to a trickle charge. For high current chargers, this cut-off voltage will be higher than for low current chargers. Therefore the manufacturer's recommended values should be followed when adjusting voltage regulators. These values will vary between manufacturers, due to the differences in rate of charge, size of batteries, and so forth.

If the gassing voltage is lower than normal, as indicated by a properly adjusted voltage regulator that fails to operate, this usually means the battery is near the end of its useful life. This is true also if the battery voltage does not let the charger give a normal low taper charge.

A lead-acid battery can be used as a fairly accurate calibrating voltage source to adjust and check D.C. voltmeters and battery tester meters in the absence of an accurate calibrating meter. Connect the D.C. voltmeter to a good, fully charged battery having a specific gravity of at least 1260, that has been standing idle for 24 to 48 hours at a temperature of about 77 degrees F. The voltage should read 2.1 volts per cell. Multiply the number of cells in the battery by 2.1 to arrive at the total voltage. For example, a 12 volt battery has six cells, or a voltage of 12.6 volts under these conditions.

Nickel-Iron Alkaline Storage Batteries (Edison Type)

The nickel-iron battery is the "granddaddy" of batteries, and is still commonly used for low current applications during long periods of time. Due to their high internal resistance, they are not suitable for vehicle starting, as is the lead-acid battery, but are used on industrial trucks and electric cars.

Their main disadvantage is initial high cost, but their many advantages offset this. Some advantages are its relatively light weight, long life, even when left in a discharged condition for long periods of time. The electrolyte must be changed after approximately 300 charge and discharge cycles. They require a special constant current charger.

Each cell of the Edison battery, as it is commonly called, uses tubes instead of plates. These tubes are perforated to hold the active elements. The positive tubes use a high nickel oxide, the negative tubes use powdered iron, and the electrolyte is a dilute solution of potassium hydrate, an alkali. These tubes are encased in a nickel plated steel case for each cell. During discharge, the chemical process changes the high nickel oxide of the positive tubes to a lower nickel oxide, and the iron of the negative tubes absorb this oxygen, changing to iron oxide. The charging process reverses this chemical action. The electrolyte composition remains the same during the charge or discharge cycle. Hydrometer readings are necessary after every 300 cycles of charge and discharge, to determine when to change the electrolyte solution. The normal specific gravity is about 1200 and decreases slowly as the cell is used. The manufacturers usually recommend that the electrolyte be replaced when the specific gravity has fallen to 1160 (usually after 8-10 months of daily service). Discharge the battery before renewing the electrolyte, and give it a 12 hour charge, at the normal rate, after renewing the electrolyte.

The nominal rating of each cell is only 1.2 volts. Unlike the lead-acid battery, the voltage rises rapidly at the beginning of the charge, decreases somewhat during the second hour, and then gradually rises during the remainder of the charge. This hump in the voltage curve is characteristic of the Edison battery, and may lead to confusion as to the battery's condition. The value and duration of this hump varies considerably, even on successive charges of the same battery. The final charging voltage, with current on, is about 1.8 volts per cell, but often ranges from 1.7 to 1.95 volts per cell. The voltage rise near the end of the charge cycle is not enough to be a very good sign of complete charge. If the extent of the previous discharge is unknown, the charge should be continued at the normal rate until the voltmeter reading has remained constant at about 1.8 volts for 30 minutes.

Vehicles using nickel-iron batteries have an ampere-hour, or watt-hour meter that shows the percentage of charge left in the battery. During the charge cycle, the meter reverses and restores the pointer to 100% automatically adding about 20% more charge than discharge.

Edison type batteries must be charged at a constant current rate. The industry specifies that this charging current be no more than 20% above normal or 20% below normal during the entire charging time. So the maximum starting charge rate is 120% of normal, and the minimum finishing charge rate is 80% of the normal rate. A typical automatic charger will provide charge rates of 108% to 87% respectively, for example. The ideal would be 100% to start and finish.

These chargers are available in dual purpose combinations for charging nickel-iron and lead-acid batteries.

Nickel-Cadmium Batteries

Nickel-cadmium batteries have high energy, and are compact, hermetically sealed, long lasting and economical. They may be charged many times. These batteries respond equally well to a trickle charge or a fast charge, and are unaffected by idle periods. The nominal voltage is 1.25 volts.

It is difficult to predict the life of a rechargeable nickel-cadmium battery "D" cell. A typical battery has a service life of only 80 hours at 50 M.A. However, this life must be multiplied by the 200 to 300 times that this type of battery can be recharged. On the other hand, recharging a nickel-cadmium cell is a critical operation, and the charging rate is determined by the capacity of the cell. Commercially available chargers are generally designed to the needs of a specific size of battery cell.

Alkaline Batteries

Alkaline energizer batteries are ideal for use where dependable high current drain and continuous service is desired for portable TV sets, flashlights and so forth. They deliver high amperage over a much longer period of time than zinc-carbon dry cells. Nominal voltage rating is 1.5 volts. Normally, they are not considered rechargeable, but can be rejuvenated a few times by placing them on a low charge, as with a zinc-carbon dry cell. The Mallory alkaline, or Durocell battery, is the only alkaline battery considered to be fully rechargeable, and only the Mallory charger should be used.

Mercury Batteries

Mercury batteries have a nominal voltage of 1.35 volts per cell. Charging of mercury batteries is not recommended, because of the danger of explosion. However, if the current is limited to a low safe value, they can be rejuvenated several times. They are designed for low to medium current drains with a flat discharge curve. The cost is relatively high, and they are not useful below 32 degrees F. Since the discharge curve is flat, and they hold a fairly constant voltage over the entire discharge period, they are widely used in calibrated instruments, such as dwell-tachometers.

Silver-Cadmium Batteries

Silver-cadmium batteries are relatively light in weight, and permit more power in a given space, reducing the size and weight of portable equipment. Their output equals the output of nickel-cadmium cells, but take only half the space. These batteries have up to five years wet, activated shelf life, with excellent charge retention. They hold the charge five to eight times longer than comparable nickel-cadmium cells, and can be recharged almost 2½ times faster. Silver-cadmium cells are rechargeable with the proper charger. They have a nominal voltage of 1.1 volts.

Carbon-Zinc Dry Cell Batteries

Carbon-zinc dry cells are made with a carbon rod in the center of a zinc container and an active chemical packed in the container. The center post or carbon rod is the positive terminal and the zinc can is the negative. They are the standard dry cell used for years in flashlights, buzzers, toys and the like. They have a nominal voltage of 1.5 volts.

They can be rejuvenated, renewed, or reclaimed, by sending 150 m.a. through one, or a group in series, for four hours, or longer. Most cells return to above normal voltage in four to five hours at 150 m.a. Let them set six hours before using to allow them to stabilize to normal voltage of 1.5 volts. This reclaiming process can be repeated from about four to thirty times. Best results are obtained from fresh, new cells run down over periods of about two weeks, reclaiming before they are entirely dead. Cells that remain in a totally discharged condition for long periods can rarely be reclaimed.

Small battery chargers are available commercially for reclaiming dry cells.

Silver-Oxide Batteries

Silver-oxide batteries, available in "button" form for electric watches and the like, have a nominal voltage of 1.5 volts per cell. They may be rejuvenated with a very low charging current.

Battery Symbols

Battery symbols, as shown in schematic wiring diagrams, consist of a series of parallel lines, drawn alternately long and short. A long and a short line constitutes one cell. Polarity is shown as a plus or minus sign with the long line being positive and the short line being negative.

TIMERS AND TIMING CIRCUITS

11

Timers used on battery chargers are of two general types—mechanical clocks, and electrically driven clocks. Both types are available with timing cycles of 15 minutes to several hours and even several days.

Mechanical Timers

Mechanical timers usually have the general appearance of a small, round can with the shaft protruding from one end and two terminals on the opposite end. They have only an "on" and "off" SPST switch, or two terminals, which turns the 115 V.A.C. off after a predetermined charging time. Small 20 second mechanical timers are used on some battery testers to time the load cycle. They are actuated by a small lever protruding through the tester panel.

Mechanical timers consist of a switch operated by a clock mechanism which is wound when the dial is turned clockwise. It should always be turned at least one fourth of the way, and then back for short timing periods. This winds the spring to start the clock. Some timers have a "hold" position to stop the clock during continuous charging. This can be tested by listening for the familiar "tick-tock" of the clock, which should be quiet on the "hold" and "off" positions, but run at all other points.

There are two common complaints or defects in the mechanical timer. Either the clock does not run, or the contacts do not close or open. Some of these timers can be disassembled and repaired. First, mark the case and the bakelite bottom. Although they can go together only one way, it saves time. Turn the clock to "off". If there are three visible screws around the base, remove them and push the assembly out by pressing on the shaft while holding the metal body, or case, with the other hand. Remove the bakelite switch. Even if the clock runs well, soak it in watchmakers solvent, or use dry cleaning solvent. It is as safe as kerosene but cleans like gasoline and evaporates. Examine the switch, and move the bakelite cam as far as it will go in the "on" position. It should not touch the contact arm or lift it open. If it does, clean the points, and bend the movable contact arm with flat-nosed pliers. Grasp the contact arm about ½-inch from the contact and lift slightly with a twist to the right. There should be good contact pressure, and the bakelite cam should clear the movable contact arm.

Check the clock to see if it will run, after blowing moisture out with low pressure air directed away from the hair spring. If the clock does not run well every time without fail or faltering, it should be replaced. Use care in reassembling the timer. The clock "on-off" switch must be at "off", and the bakelite cam must be in the off position with the switch points apart. Slide the switch onto the clock, seeing that the stud on the clock enters the hole in the cam and, at the same time, lining up the mounting screw holes. Now, carefully push the whole assembly into the case so that the staggered mounting holes and the marks line up. Replace the three screws and re-test the timer several times.

Electric Timers

Figure 11-1a and Figure 11-1b show two typical open type electric timers commonly found in battery chargers. Both have a synchronous clock motor driving a fibre or steel cam through a gear train. Some timers have only one drop-out point on the cam, whereas others have a two-stage drop-out. The high step on the cam closes contacts A, B, and C. The second step on the cam opens contacts A and B, but contacts B and C remain in contact. The third step, or the deepest notch, opens all contacts A, B and C. By studying the dial face and the circuit, you can see just how these contacts are connected. For example, for a "hold" connection, the clock motor is disconnected from the line by connecting one lead of the clock to contact A, and terminal B is connected to the line, while contact C is connected to the fan motor and charger transformer, and the cam is on the second step for "hold."

For a "slow charge," the slow charge resistor or reactor is connected

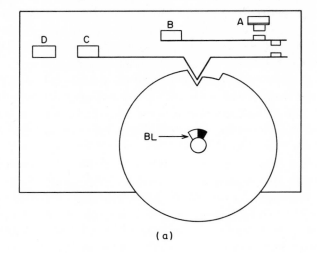

(a)

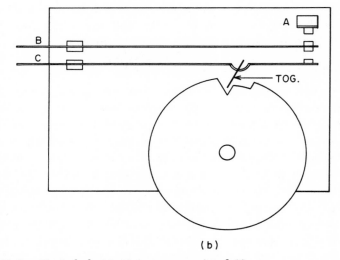

(b)

Figure 11–1   Typical electric timers, sources 4 and 12

across contacts B and C, and the cam is set on the third or deepest step. Or the "slow charge" resistor may be across contacts A and B, with the line connected to contact A, and the timer with the fan motor connected to contact C, with the transformer connected to contact B. Or the line may go to contact B, with the transformer and fan motor to contact A, and the timer motor to contact C.

These timers must have a snap action when reaching a point on the cam where it is to open a contact. It must snap open quickly instead of opening slowly. This is done in one of two ways. Figure 11-1a shows a typical timer where the contact C spring has a "V" shaped bend which rides on the cam. The cam has a loose fit on the shaft to give it a backlash as shown at BL so that it can jump backwards as the "V" comes to the edge of the cam notch, thus permitting the contact to break quickly. In Figure 11-1b, this is done by a small toggle TOG which is dragged at an angle while contacting the highest diameter of the cam. But when it reaches the drop-out point, the small toggle flips over to the opposite angle, dropping suddenly into the cam notch and opening the contacts.

Sometimes the cam wears near the drop-out point so much that it permits contacts A and B to open slowly before reaching the drop-out point. This condition is corrected by using a small, high-speed grinder to cut back on the cam about ⅛ inch, maintaining the same angle, or until the full diameter is reached. Dress off any burrs and lubricate the cam with a dab of Molycote G lubricant or any high pressure grease. Clean and adjust the contacts by bending them to get a good firm contact, yet retaining the correct spacing for positive opening and closing.

Figure 11-2 shows another open type electric timer which is similar to the timers shown in Figure 11-1 except for the contact arrangement. The identification is the same, and the description applies equally. The high step F on the cam actuates the switch for fast charging, closing contacts A, B and C, and the second step S is used for slow charge, closing C and B and opening A and B. Notch O turns the charger off, with contacts A, B and C all open. The back-lash BL arrangement is the same used as that shown in Figure 11-1a.

Three enclosed types of electric timers are shown in Figures 11-3, 11-4, and 11-5. In the open type timers the contact arrangement and cam switching can easily be seen. However, on the enclosed type, it may be necessary to "ring out" the contact circuits using the method described under Switches, Chapter 13, Section III. The dial face usually shows where the switching is to take place, so "ring out" all connections at each switching point.

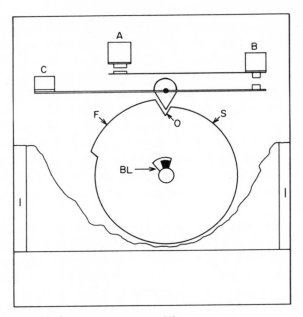

Figure 11–2   Typical electric timer, source 25

The timer shown in Figure 11-3 can have many different timing periods and sequences, although they all have the same external appearance. Some timers may operate for 60 minutes, 90 minutes, 6 hours and so forth, with a slow charge at the end of the cycle; and others may operate for 20 or 30 minutes full charge before activating a voltage regulator control circuit which takes over the control. Figure 11-3a shows the timer external connections, and Figure 11-3b shows the internal cam and switching arrangement. The two cams, CAM1 and CAM2 are driven by a synchronous electric motor through a gear train. The dotted line between terminals C and F is a connecting bar used on some models. Contacts C and B make contact before C and A, and also contacts C

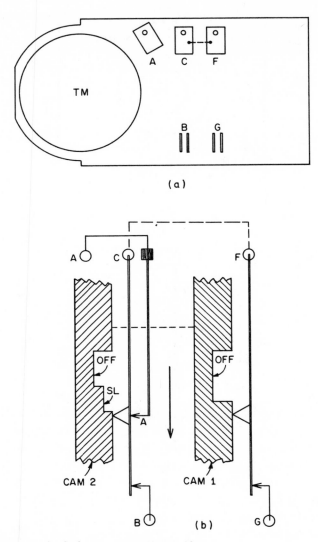

(a)

(b)

Figure 11–3   Typical electric timer, source 12

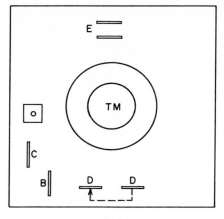

(a)

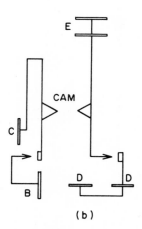

(b)

Figure 11–4   Typical electric timer, source 7

and A break contact before contacts C and B. When CAM2 is in slow charge SL position, contacts C and B are closed, and A is open. At the "Off" position contacts A, B and C are open.

The special timer shown in Figure 11-4 is a dry charge activator switch that gives the correct automatic sequence for initially charging a dry charge battery before placing it in service.

Figure 11-4a shows the external terminal connections, and Figure 11-4b shows the internal switch connections. In the "Off" position, contacts C and B are open, and contacts D and E are open. On the first cam step, contacts C and B are closed while contacts D and E remain open. On the second cam step, contacts D and E are closed while contacts C and B are open.

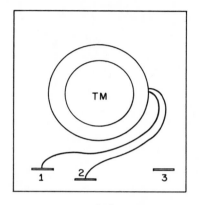

(a)

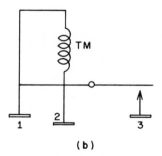

(b)

Figure 11–5   Typical electric timer, source 7

Figure 11-5 shows a simple SPST switch operated by the electric timing motor TM. Figure 11-5a shows the external connections, and Figure 11-5b shows the internal connections. The contacts between terminals 1 and 3 are normally open when timer is in the "off" position.

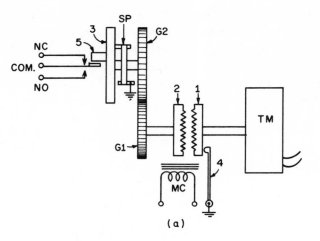

(a)

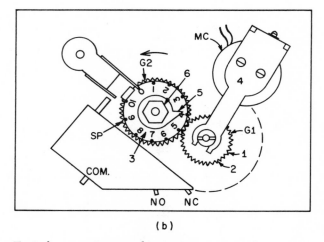

(b)

Figure 11–6   Typical automatic reset electric timer, source 19

The timer shown schematically in Figure 11-6a, and as it actually appears in Figure 11-6b, is an automatic reset type timer. The timing cycle restarts from the beginning whenever the current to the magnetic clutch coil is interrupted. The electric clock motor runs continuously. The timing cycle starts only when the magnetic clutch coil MC is energized. This pulls the armature 4 closed, engaging the clutch, pieces 1 and 2, and timing the gears G1 and G2. When the timing cycle is complete, the spring loaded switch actuator pawl 5 engages the operating arm to the micro-switch which turns the charger off. If the current is interrupted before the timing cycle is complete, the magnetic clutch coil is deenergized, releasing the clutch. Then, the spring loaded switch actuator returns to the starting position under power of the built-in spring SP. The "on" cycle can be adjusted from 0-10 hours on dial 3 by loosening nut 6 and moving pawl 5 to the desired time. It is shown on 4 hours, although this is usually set at 10 hours at the factory for golf cart chargers.

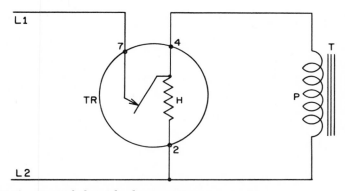

Figure 11–7   Typical thermal relay type timer, source 12

After the charger has operated long enough to actuate the TVR and start the timing cycle by operating the magnetic clutch, the timer may be tested in a few minutes without waiting for the entire cycle to pass. After the terminal voltage relay TVR has actuated the magnetic clutch, pull the A.C. plug, deenergizing the magnetic clutch coil. Move the gears around until pawl 5 is about to actuate the micro-switch. Holding the gears in this position, plug in the A.C. cord which will actuate the magnetic clutch and hold it. On late models, the gears can be turned without disconnecting the A.C. The timer then should complete the timing cycle in a few minutes. The timer is shown connected in the charging circuit in Figure 1-38, Chapter 1, Section I.

Figure 11-7 shows a thermal relay TR used as a sequence timer, as used on battery maintainers. The relay contacts 4 and 7 are closed when the heater H is cool, and open when the heater H is hot. From a cool start, A.C. line voltage energizes the charger transformer T primary P through closed contacts 4 and 7. This applies a voltage across the heater H of the relay TR. The heat operates a bi-metal thermal strip that opens contacts 4 and 7, stops the charger, and, at the same time, deenergizing the heater H. The heater, upon cooling, permits the bi-metal thermostat contacts 4 and 7 to close, and the cycle repeats over and over again. The thermal relay is designed to be "on" a certain percentage of the time. In this case, the ratio is 1 to 36. For example, it is on for 1 minute and off 36 minutes.

12
SOLENOIDS
AND RELAYS

Battery chargers and testers use many types of solenoids and relays. A solenoid is a coil with the armature being a movable plunger through the center of the coil making the contacts. A relay is a coil with an iron core which pulls a flat armature with contacts against the iron core. A.C. solenoids and relays have a low resistance "shading coil" in the iron core to prevent buzzing, or chattering, of the armature. The shading coil separates the iron core into two flux paths. The phase of the flux in the part surrounded by the shading coil is lagging the flux in the other part. Thus, there is always some flux in the iron core as the sine wave passes through zero, resulting in quiet operation.

D.C. Solenoid Construction

The most common solenoids are shown in Figure 12-1. A three terminal solenoid is shown in Figure 12-1a. The two large terminals 2 and 3 carry the heavy load current when contact 4 is engaged with contacts 2 and 3. Terminal 1 goes to one end of the coil and the other end of the coil is connected internally to terminal 2 marked BAT.

Figure 12-1b shows the same solenoid, except that the coil ends are insulated from terminals 2 and 3 and brought out to terminals 1 and 5.

Many solenoids resemble externally the two shown in Figure 12-1a and Figure 12-1b, but they may differ internally. The main difference is in the coil. Some are made for intermittent duty such as in vehicle starting solenoids and in battery testers, where they are used for only a few seconds or minutes. They also use different coil voltages. It is difficult to determine the exact application, or duty cycle, by resistance measurement alone so the voltage is usually stamped on the solenoid case. Some solenoids are also made for continuous duty, for use in battery charger circuits, for automatic controls, in alternator protectors, and in golf cart speed control switches.

Using the resistance measurements given in Table 12-1 will help determine solenoid application, and whether it is for continuous or intermittent duty. These values apply to the solenoids shown in Figure 12-1. The continuous duty type solenoid can usually replace the intermittent type solenoid, but intermittent duty type solenoid cannot replace a continuous duty type solenoid without quick failure and possible damage to other components due to the high-current draw. Table 12-1 shows some of the common solenoids in use.

Solenoid Applications and Resistances

Voltage	Duty	Coil Resistance	Application
6	Intermittent	1.2-1.5 ohms	Testers
6	Continuous	4 up	Alternator Protectors
12	Intermittent	4-6	Testers
12	Continuous	8-18	12V. Alternator Protectors
24	Continuous	16-36	Golf carts
32-36	Continuous	25-54	Golf carts

Table 12-1

These solenoids can fail in several ways. The coil can be open, grounded to the case, or even shorted. The heavy contacts can be burned, causing a voltage drop or an open circuit. A loose connection can cause the insulation at the terminal to become charred.

If the solenoid clicks on, the coil is not open. If no click is heard, check the voltage across the coil. If no voltage is present, check the alternator protector or the source of power to the solenoid coil. If the coil is defective, replace the solenoid. If the coil is good, place a heavy jumper wire across the two heavy studs. If the ammeter on the charger reads a current or the unit is functioning normally, the contact points are burned. They can be repaired economically. A voltmeter on the low range can also be used to test for any voltage drop across the contact terminals. Any voltage reading on the voltmeter means a poor contact and the solenoid should be replaced or repaired. If there is still no action, there is trouble elsewhere, such as an open circuit in the cables, circuit breaker or other parts of the secondary circuit.

Every time a solenoid operates, or is moved, the round contact washer

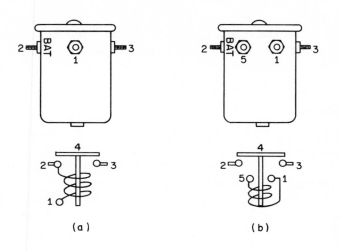

(a)   (b)

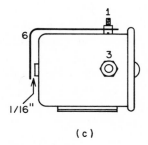

(c)

Figure 12–1   D.C. solenoid construction

may rotate, making a new contact each time. Therefore, the solenoid may check out perfectly in the shop, but the customer may bring it back, saying that it fails to charge at times. It it checks out perfectly in the shop and everything else is normal, then replace or repair the solenoid.

If the coil is good, these solenoids can be repaired economically using the proper tools. However, if warranty work is being done, the solenoid should be replaced with an exact replacement and the defective unit returned.

The solenoid shown in Figure 12-1c is similar to the one shown in Figure 12-1b except that it is modified to provide a "holding" contact. The steel strap 6 is connected to the coil terminal 1, and is bent around the solenoid case to within $\frac{1}{16}$ inch of the center core button showing externally. When the coil is energized, the magnetic flux pulls strap 6 into contact with the solenoid case. The solenoid case and strap 6 act as a switch in the circuit to hold the solenoid closed, even though the original energizing voltage is removed.

Solenoid Repair

Although the solenoids shown in Figure 12-1 are the crimped cap type, others may have a square phenolic cap that is bolted or riveted at the corners. These caps can be removed by drilling or grinding the rivets, replacing them later by machine screws and nuts.

The crimped type cap can be removed and replaced using the tools described and shown in Figure 12-2. Figure 12-2a shows how visegrip pliers can be modified to serve as an uncrimping tool, or pincers can be used. One lip is ground to a sharp edge on a radius as shown at points 1 and 2. The radius at point 2 should be the same as that of the solenoid case. Grind back at point 3 to make clearance for the coil and other terminals. To use the tool shown in Figure 12-2a, work the sharp jaw in back of the crimp and pry outward. This will bend the cap, but it should be reformed with the tools shown in Figure 12-2b.

In Figure 12-2b, piece 1 is made from solid steel stock to the dimensions shown. In the exact center of one end, drill and tap a hole for a $\frac{1}{4}$-20 machine screw. Cut off to exactly $\frac{1}{4}$ inch length and round off the end. Piece 2 is machined to the dimensions shown from a 1-$\frac{1}{4}$ inch iron pipe coupling. Piece 3 is the distorted cap to be straightened. Place cap, piece 3, on the flat end of piece 1 and tap it around the edges. Place piece 2 on an anvil and, with piece 3 on the stud end of piece 1, drive piece 1 and 3 into piece 2. Continue to pound piece 1 with a large hammer until piece 3 is back to its original shape. If necessary, use a ball peen hammer to fully restore the cap to its original curvature. Do not overdo it because too deep a cup will allow the plunger to go too far out of the coil, and will not pull in at $\frac{2}{3}$ of the rated voltage.

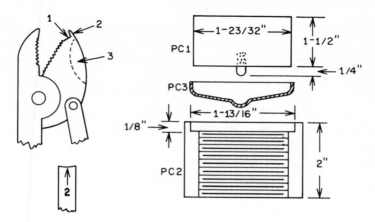

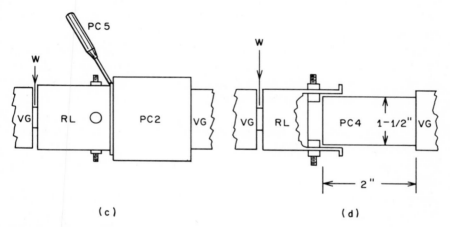

Figure 12–2   D.C. solenoid repair tools

Remove the plunger and spring from the solenoid. Disassemble the plunger and reverse the large copper contact washer, after sanding lightly. Reassemble in the reverse order.

The two heavy solenoid contacts should be removed and carefully filed clean to a flat surface without changing the angle. In most cases, these heavy contacts are reversible. In other cases, they are not the same distance from the center line of the stud and cannot be reversed without a lot of filing and shimming. Replace the large contacts and studs, and partially tighten the nuts. If they have been reversed, or there is any space between the contact and the coil, place an insulating shim in the gap to fit. To hold them in a parallel position while tightening the stud nuts, clamp the solenoid RL in a vise as shown in Figure 12-2d using tool piece 4 made of solid steel or tubing material to the dimensions shown. The two end faces of piece 4 must be smooth and parallel. A washer W with a ½ inch hole is held in position with a dab of grease. It prevents undue pressure on the protruding core stud of the solenoid case. The vise is clamped only tight enough to hold the contacts parallel while tightening the stud nuts. After removing the solenoid assemblies from the vise, insert the plunger and spring and try to rock the copper washer. There should be no rocking if the contacts are parallel.

Press the plunger with a finger. There should be additional movement of the plunger against spring pressure after the copper washer contacts the two large contacts. If necessary, change the thickness of the fibre washers on the plunger to get the correct clearance.

Connect the solenoid coil terminals to a rated voltage source, and check operation before replacing the cap. The plunger should seat completely on ⅔ of the rated voltage. Spacing and spring pressure are quite critical so it may be necessary to do a little shimming of the plunger washers.

To replace the cap, use tool piece 2 and washer W as shown in Figure 12-2c, clamping solenoid RL tightly between vise jaws VG. With a pin punch, piece 5, recrimp the cap in position in at least 3 equally spaced spots. Test the action of the solenoid on ⅔ of rated voltage. The heavy contacts should be open when the coil is not energized, and closed when the coil is energized. If the solenoid operates properly, then completely crimp the cap as shown in Figure 12-2c all around. To finish the job, seal the crimp against moisture and dust with a good cement, or wrap a piece of ½ inch × 0.007 inch plastic tape around the crimp.

Special Thermostat Controlled Solenoids

The D.C. solenoid shown in Figure 12-3a and schematically in Figure 12-3b, is used on automatic cut-off chargers. A thermostat is placed in a battery cell that cuts the charger off at a temperature of about 125

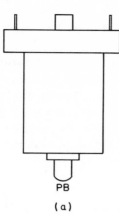

PB

(a)

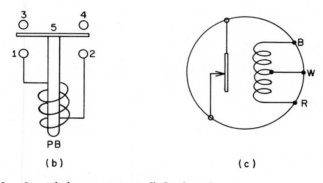

(b)   (c)

Figure 12–3   Special thermostat-controlled solenoids

degrees F. The charger must be started or restarted by pushing a "start" button PB. The coil terminals 1 and 2 are connected in series with the thermostat and across the D.C. battery clips, but the coil is not strong enough to pull the plunger and close contacts 3 and 4 by bar 5. However, when the push button PB is pushed in, the coil is energized, closing contacts 3, 4 and 5, providing that the N.C. thermostat is cool and turned on. Contacts 3, 4 and 5 are connected in series with the 115 V.A.C. power line. The coil on a typical model has about 40 ohms resistance.

The solenoid schematic shown in Figure 12-3c has the same general appearance as Figure 12-3a, but it is a dual coil, A.C. operated solenoid, using a 24 volt transformer and an N.O. thermostat placed in a battery cell. When 24 volts are applied momentarily across terminals W (sometimes blue or green) and R, the solenoid contacts close and hold to start the charger. When the N.O. thermostat reaches the cut-off temperature of about 125 degrees F., it applies 24 volts A.C. across terminals B and W opening the solenoid contacts and cutting off the charger. A momentary contact "start"-"stop" switch does the switching. In the "stop" position, the thermostat is shorted to permit turning the charger off manually.

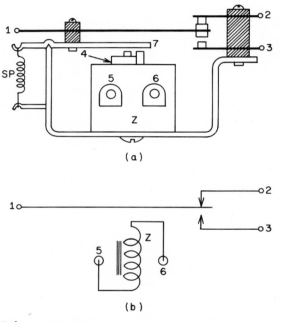

Figure 12-4   Relay construction

Relay Construction

There are many types of relays used in battery charger automatic and voltage regulator circuits, and in alternator protectors. The relay shown in Figure 12-4 is typical, but each type of relay has a different coil resistance, contact arrangement, and means of adjusting.

Figure 12-4a shows a common method of construction, and Figure 12-4b the schematic of a typical relay having a single pole, double throw switch arrangement, with one normally closed (NC) contact 2 and one normally open (NO) contact 3. The word "normally" always means the position of the switch contacts with the coil deenergized.

A.C. relays can be identified by the copper washer or "shading coil" 4 usually located on the iron core, just above the coil, or it may be hidden. Sometimes, they are used on relays operating on rectified D.C. current, or a D.C. relay may have a capacitor shunting the coil to prevent chattering.

When the proper voltage is applied across the coil Z (terminals 5 and 6), the armature arm 7 is pulled to the core against the opposing spring SP, and the flux returns through the iron core, frame and armature. The contacts are insulated from the frame and armature. Contacts 1 and 2 open and contacts 1 and 3 close when the coil is energized.

Some relays have a critical closing and opening voltage. These are used on automatic voltage selector circuits in battery chargers, testers and other dual-voltage equipment, such as ignition timing lights and the like. The relay will not operate, for example on 6 to 9 volts. If it is connected to a circuit that operates on 6 volts. Between 9 and 12 volts or higher, the relay operates, opening contacts 1 and 2, and closing contacts 1 and 3 which are connected to a circuit designed to operate on 12 volts. Once they are adjusted, they remain quite accurate over a long period of time. Relay contacts are adjusted by increasing or decreasing the spring SP tension either by bending or screwing down the spring anchor or by adjusting the contact point spacing. The coils of relays are usually marked with the voltage, AC or DC, current and resistance.

Solid state relays such as SCR, transistors are covered in other chapters.

13
SWITCHES

Many types and sizes of electric switches are used in battery chargers; among them, rotary, toggle, slide, rocker, wafer, push button and knife switches.

Rotary Switches

Figure 13-1 shows two common types of rotary switches. Figure 13-1b shows the schematic diagram which applies to both types. Figure 13-1a shows the open type. All rotary switches are connected to increase the charging rate when turned to the right or CW. Some switches are enclosed, and some even have auxiliary switches combined. Some switches have no terminal at the "off" position, or the first terminal is not connected and is used as an "off" position. Rotary tap switches are usually used as charging rate control switches, either in the primary, or secondary circuit.

The enclosed type rotary switch shown in Figure 13-1c usually is marked, such as L1, to show the common terminal or movable contact. The one shown has an N terminal connected internally to L1 for a tie-in to the fan motor. The taps are identified as 1, 2, 3 or A, B, C or may have other markings, or no markings at all. If the switch is being replaced with an identical part number, simply transfer one lead at a time.

How To Determine Enclosed Switch Connections

The procedure for determining the internal terminal connections of an enclosed switch, timer, or relay is given below. If there are no terminal markings or only a few terminals are marked, make a sketch or drawing

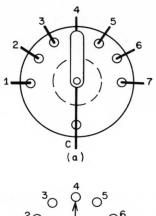

(a)

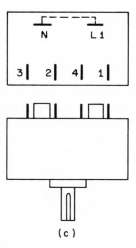

(b)

(c)

Figure 13–1   Rotary switches

of the switch. Then, make a table similar to Table 13-1, placing the terminal connections in the left-hand vertical column and the switch positions in a horizontal row across the top of the table.

Tabulation of Concealed Switch Connections

Position Terminals	1	2	3	4
L1	X	X	X	X
N	X	X	X	X
1	X			
2		X		
3			X	
4				X

Table 13-1

Refer to the sample switch shown in Figure 13-1c while doing the procedure. The sample switch has four positions, no "off" position, no stop or starting point, and six terminals marked as shown. If a switch has a stop or starting point, use the furthest left position as position 1. Use a continuity tester, such as an ohmmeter, buzzer, or test lamp. The latter two testers are slightly faster and easier to use because a short is immediately heard or seen.

1. Connect the test set between L1 and N, L1 and 1, L1 and 2, L1 and 3, and L1 and 4 in that order. A short or connection is shown between L1, N, and 1 so mark an X opposite those three terminals under position 1 on the table.

2. Connect the test set between N and 1, N and 2, N and 3, and N and 4. Terminals N and 1 are shorted together as already marked on the table.

3. Connect the test set between terminals 1 and 2, 1 and 3, and 1 and 4. None of these terminals are shorted so no mark is placed in the table.

4. Connect the test set between terminals 2 and 3 and 2 and 4. These terminals are not connected so no mark is placed in the table.

5. Connect the test set between terminals 3 and 4. These terminals are not connected so no mark is placed in the table.

6. Turn the switch clockwise to the next position which is position 2. Repeat steps 1 through 5, marking the terminals that are connected together on the table.

7. Turn the switch clockwise to the next position which is position 3. Repeat steps 1 through 5, marking the terminals that are connected together on the table.

8. Turn the switch clockwise to the next position which is position 4. Repeat steps 1 through 5, marking the terminals that are connected together on the table.

Continue the procedure until all the positions of the switch have been tested and marked. In this manner, any enclosed switch may be checked and marked correctly.

Table 13-1 shows that L1 and N are common at all positions and therefore are connected together internally. Terminal 1 is connected to L1 and N only on position 1. Terminal 2 is connected to L1 and N only on position 2. Terminal 3 is connected to L1 and N only on position 3. Terminal 4 is connected to L1 and N only on position 4. Note the straight diagonal line of X's, which indicates the sequence is correct. If the X's do not line up, the following procedure should be followed.

As an example of the assigned numbers turning out to be of improper sequence, this would show up as in Table 13-2. In this example, there are

Random Switch Sequence

Position		*1*	*2*	*3*	*4*	*5*	*6*
Terminals							
Col. 1	Col. 2						
6	1						X
3	2			X			
5	3					X	
2	4		X				
1	5		X	X	X	X	X
4	6				X		

Table 13-2

6 positions, but only 6 terminals, so that would indicate an "off" position, and 5 contact positions, with the sixth terminal being the common.

Make up a tabulation, assigning terminal numbers, and mark position 1 as a starting point. Enter the assigned terminal numbers in column 2, and mark an X at all terminals and positions that show continuity. There is no diagonal straight line of X's, so apparently the correct sequence was not selected.

In analyzing Table 13-2, the common terminal will have a row of X's in a horizontal line. In this case terminal 5, with no X in the position 1, is the common terminal and position 1 is the "off" position. Enter the correct sequence of terminals in column 1 starting with terminal 1 opposite terminal 5, column 2. Terminal 1 is the common terminal. Turn the switch CW one notch to position 2. Terminal 1 should be in contact with terminal 2 (terminal 4 in column 2). Enter terminal 2 in column 1 opposite terminal 4, column 2. Likewise enter 3 opposite 2; 4 opposite 6; 5 opposite 3; and 6 opposite 1. Column 1 now shows the correct terminal sequence, but not the correct numerical order. To correct the numerical order, make up Table 13-3, with column 1 of Table 13-2 rearranged

Corrected Switch Sequence from Table 13-2

Position		1	2	3	4	5	6
Terminals							
Col. 1	Col. 2						
1	5	Off	X	X	X	X	X
2	4		X				
3	2			X			
4	6				X		
5	3					X	
6	1						X

Table 13-3

numerically and entered in column 1 of Table 13-3 from top to bottom, carrying with it the corresponding terminal number from column 2 Table 13-2 and entering it in column 2 of Table 13-3. Also, transfer all the X's to their proper places. Table 13-3 now reads correctly for terminals in column 1, with a straight line diagonal of X's indicating proper sequence. The originally assigned terminal numbers were obviously not in the correct sequence, so they should be corrected. Change the originally assigned numbers as follows: 5 to 1, 4 to 2, 2 to 3, 6 to 4, 3 to 5, and 1 to 6.

If there is more than one switch enclosed in the same housing with multiple sections insulated from each other, as shown in Figure 13-2, make a separate tabulation for each switch section as shown in Table 13-4. One section consists of C to "off" and 1-6 inclusive, and the second section consists of terminals 7 and 8.

Toggle Switches

Toggle switches used in battery chargers have arrangements of single pole single throw (SPST), single pole double throw (SPDT), and double pole double throw (DPDT). The SPST switch may be a simple "off" and "on" switch, or it may be a momentary contact switch in which the

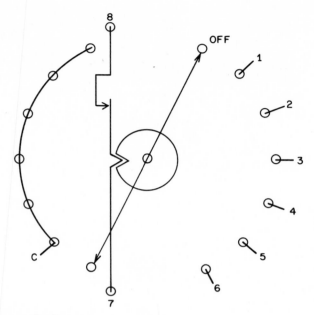

Figure 13-2   Multiple section rotary switch

toggle handle returns to the normal position when released, and may have normally open (NO) or normally closed (NC) contacts. If a switch has NO contacts, moving the toggle handle would close the contacts; whereas NC contacts would open when the toggle handle is moved. In both cases, the toggle handle would return to normal position when released. Figure

Multiple Circuit Switch Sequence

Position Terminals	Off	1	2	3	4	5	6
C	Off	X	X	X	X	X	X
1		X					
2			X				
3				X			
4					X		
5						X	
6							X
7	Off	X	X	X	X	X	X
8		X	X	X	X	X	X

Table 13-4

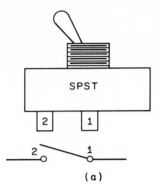

(a)

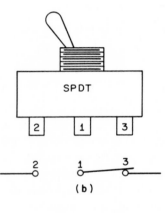

(b)

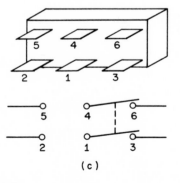

(c)

Figure 13–3   Toggle switches

13-3a shows a side view, general appearance, and schematic diagram of a toggle switch.

The SPDT toggle switch shown in Figure 13-3b, side view general appearance and schematic, may have momentary or steady contact in either or both directions, and also may have an "off" position, or no "off" position. The same can be said about the DPDT toggle switch shown in Figure 13-3c, bottom view general appearance and schematic. All of the terminals may not be used, depending on the circuit. The dotted lines between the two movable contacts indicate that they are mechanically tied together and both operate at the same time. This is true in all switch schematics.

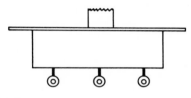

Figure 13–4   Slide switch

The DPDT switch is used to reverse polarity, by cross connecting in the form of an X one wire connected between terminals 3 and 5 and another wire connected between terminals 2 and 6.

All the switches shown in Figure 13-3 may have one of three terminal arrangements. The "quick connect", or push-on terminals are widely used in late model battery chargers. However, they may have solder connections or screw type connectors.

All switches are rated in amperes, voltage and horsepower (HP); for example, 10A-250VAC, 15A-115VAC, ¾HP-250VAC. Usually they are single-hole mounted, with a guide key. The handle usually points opposite to the live contact. For example, in Figure 13-3b, the handle is in a position closing contacts 1 and 3. In Figure 13-3a the switch is shown in the "off" position.

Slide Switches

The information on toggle switches applies to slide switches. Slide switches are used in low current circuits and carry low current ratings, such as 4A-125VAC, 2A-250VAC, 1A-125VAC. The one shown in Figure

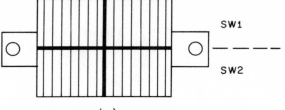

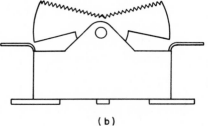

(a)

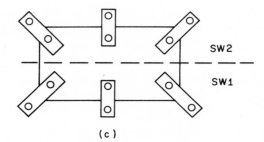

(b)

(c)

Figure 13–5   Rocker switch, source 25

13-4 is commonly used in low current chargers as a 6/12 volt switch in the transformer primary circuit, or in testers.

Rocker Switches

Rocker switches, so called because the push button action is like mounting a "V" block on a toggle switch and rocking it one way or the other, have a "rocker" action as shown in Figure 13-5. The switch shown has two separate switch sections. SW1 is a SPDT switch used to select 6 or 12 volts with an "off" position. The button of SW1 is white. Switch section SW2 is a SPDT switch with no "off" position, used for "high" and "low" charging rates, and the button is red. These switches are used on 6/12 volt, 20 ampere and smaller battery chargers. Figure 13-5a shows the front view, Figure 13-5c the bottom view, and Figure 13-5b the side view.

Wafer Switches

Wafer switches, so called because the sections are made up of thin wafer-like assemblies, may be stacked on a shaft, or operated by a handle protruding from one side. They must be limited to low currents and have very low contact resistance. They are usually used in testers and charger control circuits.

The switch, shown in Figure 13-6a physically and Figure 13-6b schematically, has three independent SPDT switches operated by one lever handle that is returned to the center position by a spring. Push the lever one way and movable contact 1 connects to contact 2 in each section SW1, SW2, and SW3. Release the lever and the switches return to center (contact 3). Push the lever in the opposite direction and movable contact 1 moves to contact 4 of switches SW1, SW2, and SW3. Release it and contact 1 returns to contact 3 of switches SW1, SW2, and SW3.

Push Button Switches

The most common types of small push button switches are shown in Figure 13-7 for low current controls.

Figure 13-7a shows a momentary contact SPST switch which may have NO contacts or NC contacts as shown schematically.

Figure 13-7b shows a momentary contact push button SPDT switch having a movable contact C and a stationary NC contact, as well as a NO contact. When the button PB is pushed, contact C is disengaged from contact NC and engages contact NO. When the button is released, it returns as shown schematically.

Figure 13-7c shows a remote momentary contact push button switch on a long cable that is used primarily on "car start" rigs.

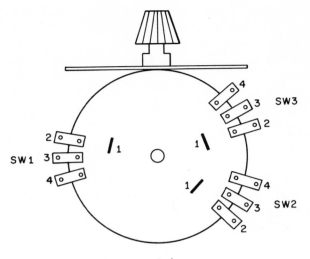

(a)

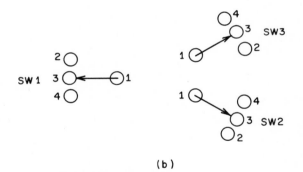

(b)

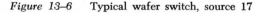

Figure 13–6   Typical wafer switch, source 17

Figure 13-8a shows a typical push button charging rate selector switch. There is usually a red or white button which is the "off" position, while the other buttons are usually black. Some selector switches do not have the "off" position. Both the front and back views show 7 buttons and 7 terminals, but the number of terminals varies. As each button is pushed,

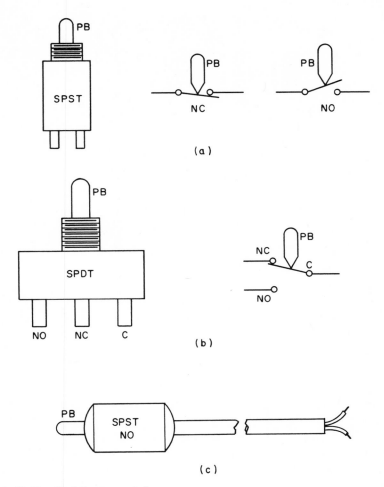

Figure 13–7   Push button switches

it pops out the previously pushed button by mechanical linkage, to prevent more than one set of contacts from closing at the same time. It is a common fault in the push button, as well as the enclosed rotary type switch, to have more than one contact engaged with the common contact. This shows up as a short across the primary transformer taps.

    A schematic wiring diagram may show the push button switch as either schematic Figure 13-8b or Figure 13-8c. Sometimes other terminals next

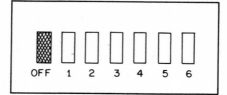

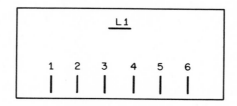

(a)

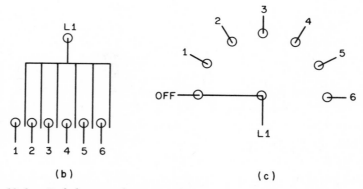

(b)   (c)

Figure 13–8   Push button selector switch

to L1 are found, which are usually connected internally to L1, but each should be checked unless replacing the switch with an exact replacement.

To check the connections of unknown, or unmarked, switches, use the same procedure described under "How to Determine Enclosed Switch Connections" earlier in this chapter.

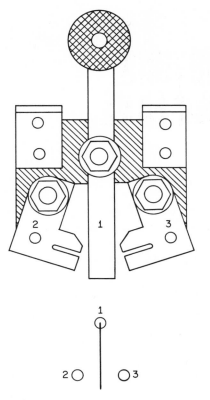

Figure 13–9   Knife switch

Knife Switches

Knife switches are not commonly used on late model battery chargers, but Figure 13-9 shows a SPDT knife switch used on a dual-voltage battery charger in the heavy current D.C. output circuit.

If this type of switch is defective and cannot be repaired, and a replacement unit is not available, an inexpensive, trouble-free substitute can be made as follows. Bring out, through the case, the two cables connected to terminals 2 and 3, and terminate the ends with a copper lug, or piece of copper tubing. Also bring out the cable from terminal 1 and terminate in a heavy battery clamp. To select 6 or 12 volts, for example, simply connect the battery clip to the correct cable terminal 2 or 3. Mark the terminals 2 and 3 for the correct voltage. Secure the cables with strain reliefs.

14
MOTORS
AND FANS

The operational life of all electrical equipment is affected by excessive heat. Equipment parts fail more quickly and the electrical output is reduced. The heat produced by circuit components which use current must be exhausted out of the battery charger unit and cool air drawn inside. Most fast-rate battery chargers, and some slow-rate battery chargers, use a small electric fan to blow air over rectifiers, heat sinks, the power transformer, and the fan motor. The electric fan consists of a 115 volt AC shaded-pole motor with a set of fan blades. Some older model battery chargers use a small 6 or 12 volt DC motor (similar to an automotive heater blower motor) that uses the DC output of the battery charger. The motor may have a series-connected resistor to limit the motor's operating voltage to the proper range. These DC fan motors are easily replaced by connecting an AC motor across the 115 volt line.

Natural convection cooling (without a fan) is sufficient for electrical equipment where heat is not excessive. But, in electrical equipment where the current demands and the resultant heat are excessive, cooling must be achieved by a fan. Some battery chargers have a fan mounted so the hot air is blown upwards but this method tends to draw dust and dirt from the floor below. Most battery chargers direct the fan either downwards or outside the case to avoid the dust problem. Compact or flat-cased battery chargers blow cool air horizontally through the case.

All battery chargers tend to collect dust and dirt which prevent proper heat dissipation and work in between bearing surfaces, switches, relays and thermostats, thus increasing wear. Therefore, every time a battery charger unit is serviced, the case should be opened and the dust and

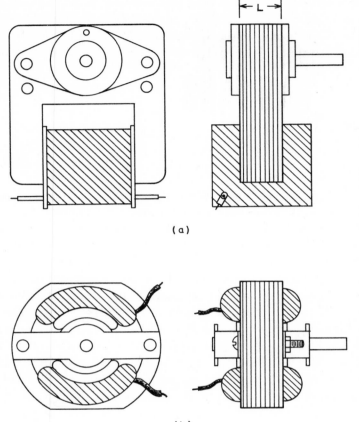

(a)

(b)

Figure 14–1   Open-skeleton motor construction

dirt cleaned out. Use a vacuum cleaner, clean rags, and a suitable non-flammable solvent to wipe all the parts clean.

Fan Motor Construction

Figure 14-1 and Figure 14-2 show the physical appearance of the most common types of fan motors used in battery chargers. They are all of the shaded pole type, but have different power outputs, different size fan blades (from 4 inches to 10 inches), mounting dimensions, and shaft diameters. Standard rotation is CW when looking at the shaft end. However, motor rotation may be CCW in some older battery chargers.

A shaded pole motor, as shown in Figure 14-3a, consists of a coil C on the stator and a squirrel cage rotor R. They are made self-starting by

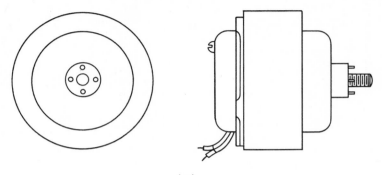

(a)

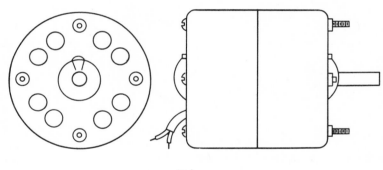

(b)

Figure 14–2   Shell-type motor construction

adding two small shading coils SC in the stator core using one turn of heavy wire. This shifts the phase of the flux in one part of the stator behind the main flux, creating a rotating field. The rotation is from the main, larger flux path, toward the shading coil SC as shown by the arrow.

To reverse motor rotation, when necessary or possible, tear down and

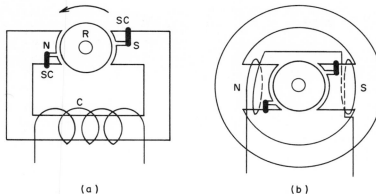

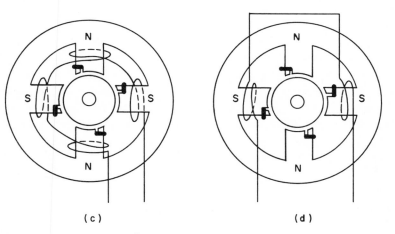

Figure 14–3   Fan motor windings

reassemble with the shaft facing in the opposite direction. Be sure that the laminations of the rotor line up with the laminations of the stator. The rotor must be centered magnetically.

Most of these small motors are used for "air over" applications, or fan duty operation. If operated without a fan, they would overheat.

Fan motors should be cleaned and oiled every time a charger is serviced. Even new motors should be oiled before using them. There is a small oil hole on most of them through which the oil is soaked up by a felt wick. There are available small oilers with a flexible spout, but a used, discarded hypodermic needle makes a good oiler, the larger and longer the needle the better. The open type skeleton motor can best be oiled if the shaft is in a vertical position so it can be oiled from both the inside as well as the outside of the bearing. Use No. 10 or 20 weight motor oil.

If the bearings are worn, and there is any side play, replace the motor, since it is too expensive to repair them, even if the bearings and rotors are available.

Although some motors have common mounting dimensions, their shaft diameters vary from one manufacturer to another. Standard shaft diameters are normally $\frac{1}{8}$ inch, $\frac{3}{16}$ inch, and $\frac{1}{4}$ inch, but some manufacturers use odd sizes such as 0.182 inch and $\frac{7}{32}$ inch diameter.

A large stock of fan blades in all sizes and bores allows many substitutions of motor and fan when the old fan has a different bore from that of the motor in stock. It is almost impossible to drill out a fan blade bore without throwing it out of balance. Many fan blades need replacing due to cracking, corrosion, loose hub, bending, and other defects.

Fan blades will vibrate and cause early motor failure if the blades do not run true. They can be aligned by holding the finger, or a tool, steadily at a point just touching the blades as they are rotated and bending the blades that are out of alignment. If bent too badly, they should be replaced. Always wipe the blades clean.

The power output of the motor varies in proportion to the thickness of the stator, or "stacking length" L, as shown in Figure 14-1a, if the other dimensions are the same. Always replace a motor with one having the same or longer stacking length.

If the motor runs slower than normal after cleaning and oiling, it should be replaced. The rotor shaft may be roughened, and would not show side play, but would cause friction. If the fan motor does not come up to full speed, check the fan blade size. Too large a fan blade will load the motor down so that it cannot reach full speed.

The unit bearing type of motor shown in Figure 14-2a has the longest bearing life compared with any of the others. Usually, there is no oil hole in the case, but if there is it is plugged with a screw on the back cover. These motors are oiled every five years by removing the back cover, or drilling a small hole on the high side of the back, adding about a tablespoon of Number 10 or 20 oil, and plugging the hole with a sheet-metal screw and fibre washer.

The open skeleton frame motor shown in Figure 14-1a is a 2-pole motor having a full load speed under 3000 RPM.

Figure 14-1b shows an open skeleton frame motor that may have 4 poles and run at about 1500 RPM, or it may have 2 poles and run at about 3000 RPM.

Figure 14-2a shows a totally enclosed unit bearing shaded-pole motor, having either 4 poles or 2 poles.

Figure 14-2b shows the shell type shaded-pole motor that may be either 2 poles or 4 poles, and may be open or totally enclosed.

A 4-pole motor is quieter in operation, but it requires a larger fan blade than a 2-pole motor to move the same amount of air.

The propeller type blade is the most popular type fan. However, the squirrel-cage type blower has been used on the older fast rate chargers because of its greater efficiency.

Fan Motor Windings

Figure 14-3 shows the various windings found in battery charger fan motors.

Figure 14-3a shows the open skeleton frame shaded pole motor, having one coil C, 2 poles [one north (N) and one south (S)], a rotor R, and 2 shading coils SC. The rotation is toward the shading coil SC on each pole, as indicated by the CCW arrow.

Figure 14-3b shows a 2-coil, 2-pole shaded-pole motor where the two coils are wound in opposite directions to create 2 magnetic poles, N and S.

Figure 14-3c shows a 4-coil, 4-pole, shaded-pole motor where the 4 coils are wound alternately in opposite directions to create 4 magnetic poles, N, S, N, and S.

Figure 14-3d shows a 2-coil, 4-pole shaded-pole motor where the 2 coils are wound in the same direction to create 2 S poles opposite each other. These are called salient poles. The return path of the magnetic flux creates 2 consequent magnetic N poles opposite each other. This gives a 4-pole motor which will operate at the same speed as the 4-pole motor shown in Figure 14-3c.

The squirrel-cage rotor is the same on all induction motors.

Heavy duty industrial chargers that operate from 230 volt A.C. power lines frequently use a standard 115 volt fan motor by connecting a series resistor in the voltage input line to give 115 volts drop across the resistor, and 115 volts across the motor.

CIRCUIT BREAKERS AND FUSES

15

Circuit breakers and fuses protect various components in a battery charger from damage by overload current, by temperature, or both. The basic element of most circuit breakers is a bi-metal strip, or disc, made by fusing two dissimilar metals together, each having a different rate of expansion from the other. When heated, one metal strip expands more than the other causing the strip to curl, or with the disc, causes it to pop like an oil can. One contact is attached to the bi-metal element and the other contact is stationary. With the disc type there are two contacts, diametrically opposite, attached to the disc, and two stationary contacts, providing a double break. When cool, it goes back to its original shape, except on the manual reset types which must be reset by pushing a button after it cools.

Low Voltage Circuit Breakers

The type shown in Figure 15-1a is a low voltage circuit breaker used in the 6 or 12 volt D.C. side of a battery charger to protect the transformer, rectifiers and meter. It is available in ratings up to 50 amperes, and is often connected in parallel with other circuit breakers to double its rating. Parallel connected circuit breakers may not always break the circuit together, but if one opens, all the current goes through the other one, tripping it immediately. Practically speaking they open at the same time.

Circuit breakers are available with 10-32 stud terminals, quick connect slip-on terminals, or both. The amperes and volts ratings are usually stamped on the circuit breaker case.

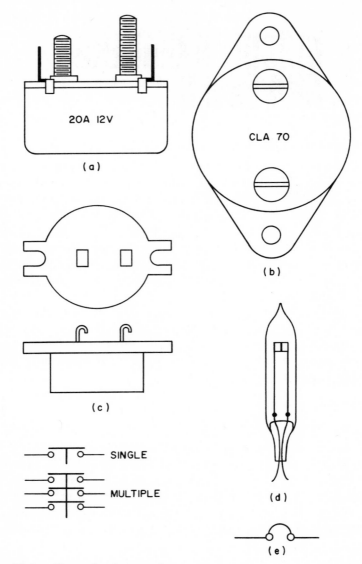

Figure 15–1   Circuit breaker switches

Figure 15-1b shows a heavy duty circuit breaker rated for 40 amperes and above, with screw-on terminals. The ampere rating is part of the catalog number. For example, the Klixon CLA-70 has a 70 ampere rating. This circuit breaker is used in the D.C. output of battery chargers at low voltages.

Figure 15-1d shows a small glass enclosed circuit breaker limited to low currents and low voltage applications, such as, trickle chargers and alternator protector circuits.

Line Voltage Circuit Breakers

The Klixon type circuit breaker shown in Figure 15-1c opens the transformer primary circuit of 115 V.A.C. when the current exceeds the rating of the breaker or the temperature reaches a dangerous value. Sometimes, these circuit breakers are mounted on the rectifier plate to give double protection from high primary current and high rectifier temperatures.

On some battery charger-testers, a circuit breaker may be found near the load, or discharge, resistor that opens the solenoid circuit if left on too long. This operates strictly as a thermostat to disconnect the solenoid by opening the solenoid coil circuit.

There are many other small circuit breakers and thermostats that protect more costly components.

Most circuit breakers are automatically reset. However, some are manually reset and require that a reset button be pushed, after cooling to restore operation of the charger.

Fuses

A fuse is a length of wire that melts and breaks apart when current that exceeds its limits is applied to the wire. Fuses are made in many different sizes and shapes. Most fuses are made by encasing the wire in a glass cylinder with metal caps on each end. The cylinder may be made of some other non-combustible material, also. The rated limits of the fuse are stamped on the metal caps. Special fuse wire links are used to protect individual components and are soldered in the circuit wiring. They are replaced by soldering a new wire link in place or by installing a new fuse in the holder. Some fuses are designed to withstand current overloads for a very short time without melting. These fuses, called "slow blow" or "Fusetron", are used to reduce the frequency of fuse replacement and still protect the equipment. In operating certain controls of the battery charger, a careless operator might turn the charge rate control or the coarse control up too fast or too high. If he turns the control back quickly enough, the "slow blow" or "Fusetron" fuse might hold. An ordinary fuse would open immediately. For example, a 6-ampere battery charger would be better protected by a 7 or 8 ampere "Fusetron" slow blow fuse than a 15 ampere fuse that might open immediately.

Fuses are installed in holders of various types, the more common being the screw-in and the snap-in styles. Fuse wire links may be installed in break-apart holders wired in the circuit wiring, otherwise they are soldered in the circuit.

Testing Circuit Breakers and Fuses

Circuit breakers can be checked for continuity by reading any voltage drop across the unit when under load. A circuit breaker that has opened many times may have burned or overheated contacts, that, although they show continuity, may have a voltage drop under load. A circuit breaker that has a measurable voltage drop, as measured on the lowest volt meter range should be replaced, because this voltage drop generates heat which will cause the circuit breaker to open prematurely at a low current level.

Fuses can be checked for continuity using an ohmmeter. Fuses are either burned open or are still good. Always replace burned open fuses with the proper size fuse after determining the cause of the fuse failure.

16 ALTERNATOR PROTECTORS

Alternator protectors are reverse polarity sensor devices that warn either by light signal, buzzer, or meter when the charger is connected incorrectly to the battery. Sometimes, they actually prevent the operation of the charger until the correct polarity connection is made.

If a charger is connected in reverse to alternator equipped vehicles, it is almost certain to destroy the diodes. Instructions say to disconnect one battery cable on alternator equipped vehicles. The only reason for doing this is because of careless operators who might get the polarity reversed. If certain that the polarity is correct, it is not necessary to remove one cable from the battery.

Polarity Indicators

The warning type of polarity indicator is shown in Figures 16-1 and 16-2. After connecting one battery clamp, it shows by a light bulb whether or not to connect the other clamp, or reverse it. Figures 16-1 and 16-2 are just the opposite in their indication.

Figure 16-1a shows the physical appearance and Figure 16-1b shows the schematic layout. A rubber holder contains a diode D and bulb B (#51 for maximum light or a #57 for longer life and lower brilliancy). Terminal A is a sharp tack pressed into the positive clamp cable to make a connection. Terminal C is about one foot long with a tip on the end. To test for correct polarity, connect the positive clamp to the positive battery post, and touch terminal C to the other battery post. ("When you see the light you know you're right"). If it does not light, reverse the positive clamp to the other battery post and test again. The polarity tester should light. If it does not light, check for a burned out bulb or an open diode. If the tester burns when connected for either polarity, check for a shorted diode.

Polarity Indicator

Figure 16-2a, as shown in a cut-away view, and Figure 16-2b shown schematically, illustrate a circuit that uses no diode, but has twin bulbs

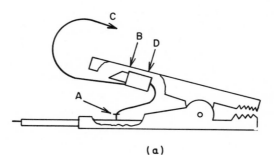

(a)

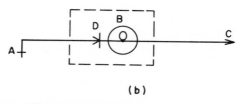

(b)

Figure 16–1   Polarity indicator, source 4

in parallel. If one bulb burns out, the other light will still glow. This circuit depends on a good rectifier when used in a charger. The polarity is right when the lights are out and wrong when they glow. This is called "stop lite", meaning stop when the red light comes on. Connect the negative clip to the negative battery post and touch the button protruding from the handle (terminal C) of the positive clip to the other post. An unlit bulb means polarity is right, because the charger rectifier will not pass enough current in the reverse direction to light the bulbs. If they do light, the rectifier is shorted. On the wrong polarity, the rectifier in the charger has a low resistance in the forward direction, and the bulbs will light.

In Figure 16-2, A is a tack on the end of the lead that is pressed into the positive cable to make contact. Bulbs B1 and B2 are #51 lamps

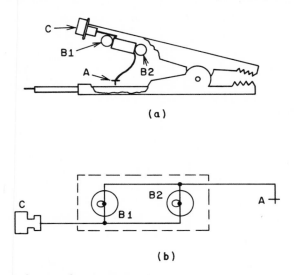

(a)

(b)

Figure 16–2   Polarity indicator, source 4

mounted in a rubber holder which is secured by sliding one of the clamp spring tangs into a hole provided for the purpose. Piece C is an insulated button, clamped in the positive cable clamp handle, used to contact the positive battery post.

Polarity Indicating Meter

Another polarity indicator uses a D.C. voltmeter connected across the D.C. output of the charger. The meter has zero voltage at the middle of the scale. With the charger connected properly to the correct battery terminals, the meter reads to the right in a green, or "go ahead" area. If the polarity of the battery is reversed, the meter reads to the left in a red, or "danger" area. Some of these meters have a voltage scale in the green area to show when the battery has reached full charge. With this type of indicator, it is necessary to have a D.C. circuit breaker to protect the rectifier on reverse polarity.

Transistorless Alternator Protector

Figure 16-3 illustrates one of the simplest alternator protectors, or reverse polarity protectors. It uses only two bulbs, one small selenium rectifier or silicon diode, and a solenoid for control.

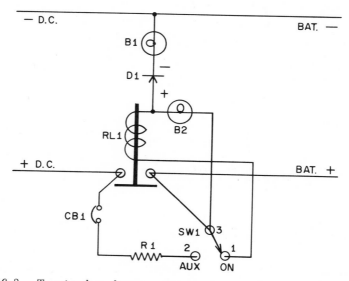

Figure 16-3   Transistorless alternator protector, source 17

Switch SW1 is a spring-loaded, push button switch that is normally in the "on" position. When it is depressed to the "aux" (auxiliary, or dead battery) position, it has to be held down. When released, it returns to the "on" position.

With the battery connected and the charger on, the positive battery voltage goes through switch SW1 contacts 3 and 1, through the coil of solenoid RL1, rectifier D1 and ballast lamp B1 (#1073), thereby closing the solenoid contacts and connecting the charger to the battery. Bulb B2 (#1815) is a pilot light that glows when there is voltage across the solenoid RL1 coil. This coil voltage is about 5 volts when charging a 6 volt battery, and about 6.5 volts when charging a 12 volt battery. This is due to the low cold resistance characteristics of the ballast lamp B1. Both bulbs will light when the polarity is correct; neither will light when it is incorrect.

If the battery has such a low voltage, or no voltage, that it cannot operate the solenoid RL1, and the polarity is correct, then press the switch SW1 to close contacts 3 and 2. This disconnects contacts 3 and 1 and the solenoid RL1, and places a small charge of 3.5 to 5 amperes across the battery until the voltage can operate the solenoid when switch SW1 is released, again establishing a circuit between contacts 3 and 1. Resistor R1 is a 2 ohm, 25 watt resistor used to limit the charging current. The charger rate control should be on a low range. Circuit breaker CB1 will cut out if overloaded due to a shorted battery or reverse polarity.

If the battery polarity is reversed, the rectifier D1 will not pass current to the solenoid and there will be no action, neither bulb will light, and the solenoid will not operate. The rectifier D1 is not critical, and can be replaced, if defective, by a selenium or silicon unit rated at 2 amperes or more. It uses convection cooling. Usually, a rectifier is used in fan cooled chargers, so it should run cool, especially if mounted on a heat sink.

The ballast lamp B1 is critical and should be replaced with only a #1073 bulb. Pilot lamp B2 is not critical and can be replaced with any bulb drawing no more than 0.20 amperes at 12-16 volts and having the same base.

Solenoid RL1 is a 6-volt, 4-ohm continuous duty, 4-post unit.

Caution: On most alternator protectors, do not reverse the cable clamps while the charger is on after the solenoid closes on the correct polarity. Reversing the clips leaves no protection at all because the charger voltage holds the solenoid contacts closed. To properly check the reverse polarity action of the alternator protector, disconnect the 115 VAC to deactivate the solenoid, then reverse the clips and reconnect the AC. Most alternator protectors will show the same indication on an open D.C. circuit, such as a poor connection, as it will on reversed polarity.

Alternator Protector

The alternator protector shown in Figure 16-4 uses a transistor, 3 bulbs, 2 resistors, a solenoid, and a small silicon diode for protection.

With the charger turned on and the battery clips disconnected, the indicator pilot light B3 burns. Resistor R1 holds the base B of the transistor Q1 at the same potential as the emitter E, preventing conduction between emitter E and collector C. The current goes from the charger positive terminal through B3, R2, B2 and RL1, back to the charger negative terminal. The solenoid does not close because the current through the B3 circuit is too low. When the battery is correctly connected with

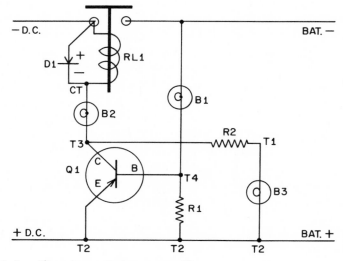

Figure 16–4   Alternator protector, source 12

good contact and no open circuit, the indicator pilot light B3 does not burn. The transistor base B is now negatively biased through base lamp B1, drawing current through the base B and emitter E of transistor Q1. This causes transistor Q1 to conduct between emitter E and collector C, carrying current from the charger positive terminal, through transistor emitter E, collector C, collector lamp B2, solenoid coil RL1, and back to the charger negative terminal. This closes the solenoid contacts, and the charger operates normally. The pilot light goes out because, while the transistor conducts, there is not enough voltage drop across the emitter E and collector C of the transistor to light lamp B3.

The small diode D1 is connected across the solenoid RL1 coil in reverse polarity and has only one function—to discharge the relatively high voltage of opposite polarity that appears across the coil when the current is interrupted. When current flows through the coil in one direction, it builds up a flux; and when it is disconnected, the flux collapses and develops a high voltage for a split second in the opposite direction.

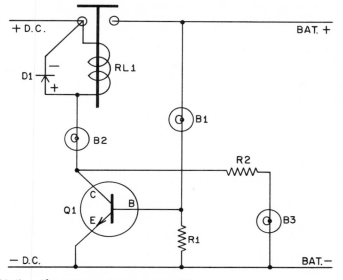

Figure 16–5   Alternator protector, source 12

The diode is the correct polarity to conduct this current and dissipate it so that it will not damage the transistor. If the transistor requires replacing, always check this diode by unsoldering one lead or otherwise disconnecting it. If it is shorted, the diode may not allow the solenoid to close. A shorted diode is usually indicated by an increase in the brightness of collector, or ballast lamp, B2.

If the battery is connected in reverse, the base B of transistor Q1 receives a positive bias, and the transistor does not conduct. Nothing happens, except pilot light B3 lights, just as in an open D.C. circuit.

When pilot light B3 glows, it means that one or more things are wrong: 1. Poor clamp connection on battery post; 2. open circuit in the D.C. cables or connections; 3. burned out base bulb B1; 4. reversed connection to the battery. Of course, there are other troubles in the alternator protector itself that could cause lamp B3 to burn, such as a defective transistor, an open base B, or a short between base B and emitter E.

Solenoid RL1 is a 6-volt, 4-ohm, 3-post unit. The indicator pilot light B3 is a sealed bulb and holder with leads. B2 is a #1073 bulb; B1 is a #433 bulb; transistor Q1 is a PNP type 2N307A; resistor R1 is 470 ohms, ½ watt; R2 is 47 ohms (originally 18 ohms), ½ watt; D1 is a small silicon diode rated about 1 ampere, 400PIV.

Alternator Protector

The circuit shown in Figure 16-5 is the same as Figure 16-4 and all components are the same, except that a type 2N3055 transistor Q1 which is an NPN type replaces the PNP type shown in Figure 16-4. Note that this requires a reversal of all polarities. Some models use a #433 bulb for B1, others use a #305 bulb.

Alternator Protector

Figure 16-6 shows a simple alternator protector having a built-in thermostat, or circuit breaker TH1. With the charger on and the battery clips properly connected, the current from the battery goes from the positive terminal, through resistor R2 (180 ohms), bulb B1 (#53), shunt resistor R3 (120 ohms), and back to the battery negative terminal. Because bulb B1 resistance is lower than R2, it places a negative bias on the base B of transistor Q1 (2N555 PNP) causing it to conduct from emitter E to collector C through ballast resistor R1, thermostat TH1 (185 deg. F), solenoid RL1 (6V, 4 ohm), and back to the battery negative. The contacts of solenoid RL1 are closed, and the charger operates normally.

Thermostat TH1 is a small, inexpensive device attached to the rectifier heat sink, which interrupts the current flow through the solenoid coil when the rectifier temperature reaches 185 degrees F. This opens the solenoid contacts and closes them again after the rectifier has cooled to a lower temperature. Ballast resistor R1 has cold, low resistance characteristics similar to those of the #1073 bulb used in other dual voltage alternator protectors.

If the polarity of the battery charger clips is reversed, the transistor base B would be positive biased, and would not conduct to close solenoid

RL1. If there is poor contact or an open in the D.C. circuit, resistor R2 would hold the transistor Q1 base B at the same potential as the emitter E, the transistor would not conduct to close the solenoid contacts, and the charger would not be connected to the battery.

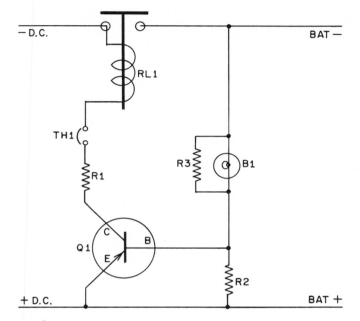

Figure 16–6   Alternator protector, source 13

Alternator Protector

Figure 16-7 shows an alternator protector using a silicon controlled rectifier SCR and three circuit breakers. Prod TH2 is a normally closed NC thermostat that is inserted in one of the battery cells, and opens at a temperature of approximately 125 degrees F., disconnecting the charger from the battery. Thermostat TH1 is a NC device mounted on the rectifier heat sink that opens the solenoid RL1 contacts at 185 degrees F. When switch SW1 (NO, shown in off position) is pressed to "start", current flows from the positive battery, through R3 (470 ohms), R2 (470 ohms), and back to the negative battery. A positive voltage equal to about half the battery voltage is tapped off at the junction of R2 and R3 and fed to the gate G of SCR (GEC106Y2). This is enough to make the SCR conduct from anode A to cathode C. Current flows from the charger positive terminal, through solenoid RL1 coil, R1 ballast resistor,

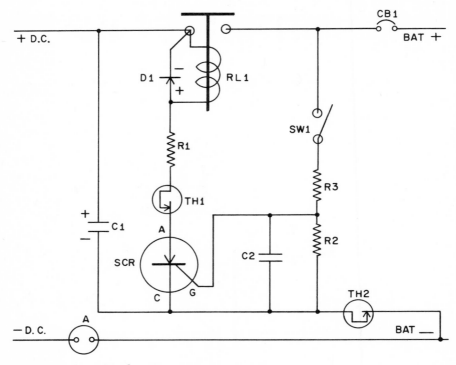

Figure 16–7   SCR alternator protector, source 13

TH1, SCR anode A and cathode C, TH2, and back to the common negative. This causes solenoid RL1 to close the contacts, and the charger works normally. Diode D1 is connected in reverse polarity and has no function but to dissipate the high discharge voltage of the solenoid coil when the current is disconnected. Capacitor C2 is a 0.47-1 MFD, 10 volt unit that steadies the voltage across the gate G and cathode C of the SCR. Capacitor C1 (200MFD, 25V) is very important as a filter to prevent the rectified current from dropping too low during each cycle. Otherwise, the SCR would stop conducting, since the gate G is only connected to start the operation. Failure of capacitor C1 will cause the solenoid to open as soon as the "start" button SW1 is released. A is an ammeter; solenoid RL1 is a 6-volt, 4-ohm continuous device having 3 terminals.

When thermostat TH2 cools down, and the contacts close, the charger will not start again until the "start" button SW1 is pushed.

On reversed polarity, nothing happens because the SCR will not conduct.

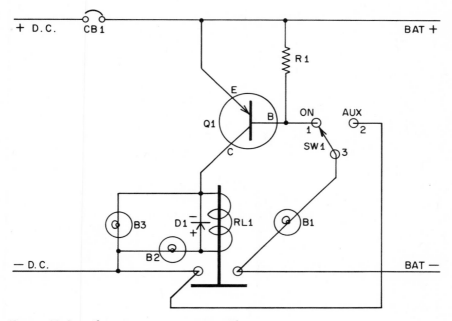

Figure 16–8   Alternator protector, source 17

Alternator Protector

In Figure 16-8, with the charger and battery connected properly, the transistor Q1 (2N441 PNP) has a negative bias on the base B, causing emitter E and collector C to conduct. This bias comes from the battery negative terminal, through bulb B1 (#313), SW1 contacts 3 and 1, to Q1 base B, and through resistor R1 to the battery positive terminal. With transistor Q1 conducting between collector C and emitter E, current flows from the charger negative terminal, through bulb B2, solenoid coil RL1, transistor Q1 collector C and emitter E, and back to the battery positive terminal. This closes solenoid RL1 contacts, and the charger operates normally. Pilot light B3 (1815) is across the series connected solenoid coil and ballast lamp B2. Pilot light B3 will glow when the polarity is correct, but will be dark on reversed polarity, or if the D.C. circuit is open, or if bulb B1 is burned out. If the battery is dead, or the voltage is too low to operate the solenoid when properly connected, push

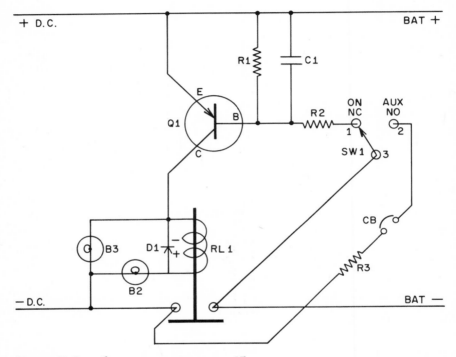

Figure 16–9   Alternator protector, source 17

switch SW1 to AUX (auxiliary) position to close a circuit across the solenoid contacts through bulb B1. This allows a small charging current to enter the battery and build the voltage up enough in a few seconds so that when SW1 is released it should close the solenoid contacts. Switch SW1 is a momentary contact switch with contacts 1 and 3 NC in the ON position. D1 is a small diode that discharges the solenoid coil when disconnected.

Alternator Protector

The alternator protector shown in Figure 16-9 is essentially the same as in Figure 16-8, except that it uses an auxiliary charging resistor R3 and circuit breaker CB, instead of bulb B1, and uses resistor R2 instead of base lamp B1 to limit the bias on the base B of transistor Q1. The pilot light B3 glows on correct polarity, but is dark on reversed polarity. The

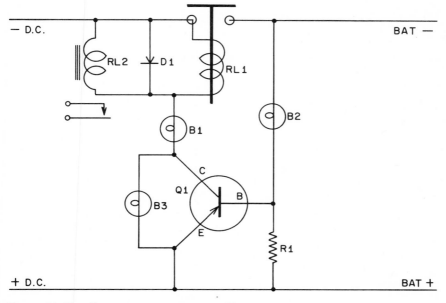

Figure 16–10   Alternator protector, source 17

pilot light is on the front panel behind a green lens on most of these models that have the pilot light connected across the solenoid. Those that have the pilot light across the transistor collector C and emitter E will have a red pilot light. The condenser C1 is used only in some models to steady the bias voltage.

Alternator Protector

In Figure 16-10, when the charger is connected to 115 VAC, and the battery is properly connected, a negative bias is placed on the transistor Q1 base B through base lamp B2 (#313), and resistor R1 (100 ohms).

Transistor Q1 conducts through emitter E and collector C, ballast lamp B1 (#1073) and solenoid RL1 coil, closing the contacts of the solenoid RL1. At the same time, relay RL2 contacts close the primary circuit, and the charger operates normally. Pilot light B3 (#53) goes out when transistor Q1 is conducting because there is very little voltage drop across the transistor collector C and emitter E. Relay RL2 coil is connected in parallel across the solenoid RL1 coil so both operate at the same time, closing a switch in the 115 VAC primary circuit.

Multi-Voltage Alternator Protector

Figure 16-11 shows an alternator protector having its own D.C. power supply to operate the control circuit. This circuit, or one similar to it, is

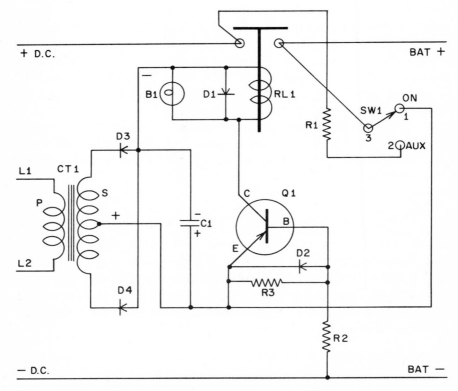

Figure 16–11   Multi-voltage alternator protector, source 17

usually used on chargers providing a wide range of battery voltages such as, 6/12/18/24 volts. This is too wide a range to handle by ballast bulbs or resistors. There is a constant voltage to the alternator protector, regardless of the battery voltage.

Here, the negative bias for the base B of transistor Q1 is obtained in the usual manner across the battery, from the battery positive, through switch SW1 contacts 3 and 1, R3, R2, and back to the battery negative, with the junction of R2 and R3 connected to base B of the transistor Q1. This makes transistor Q1 conduct from the emitter E to collector C. The separate, isolated D.C. power supply now supplies current through the transistor collector C and emitter E, to the solenoid closing the contacts, and allowing the charger to operate normally. Pilot light B1 (#1815) is connected in parallel with the solenoid coil and glows when polarity is right. Diode D1 is the solenoid coil discharge diode. The separate power supply consists of a control transformer CT1, which is a center-tapped, full-wave circuit with diodes D3 and D4 and filtered capacitor C1. If the battery clips are reversed, the transistor base B has a positive bias, or no bias, and will not conduct. Diode D2 is a small diode (same as D1) connected across resistor R3, that holds the base B voltage at about the same voltage as emitter E on reversed polarity.

If the battery voltage is too low to operate the solenoid, press switch SW1 to the AUX position, which connects a charging resistor R1 across the solenoid contacts to charge the battery for a few seconds, so that when SW1 is released, there is sufficient battery voltage to operate the solenoid.

Single-Voltage Alternator Protector

Figure 16-12 shows a simpler alternator protector designed for 12 volts only. Transistor Q1 (2N441) has a negative bias on base B through bulb B1 (#1445). The transistor conducts through emitter E and collector C, and solenoid RL1, closing the solenoid contacts for normal operation. A push button switch SW1 is provided where the alternator protector is used for "car start" or "hot boxes".

Transistorless Alternator Protector

The alternator protector shown in Figure 16-13 uses two relays to operate a dual-voltage 6/12 volt unit. Relay RL1 is a voltage-sensitive relay having a SPDT switch which opens contacts 3 and 1 and closes contacts 3 and 2 at about 9 volts or over. This acts as an automatic voltage selector.

With the charger connected and the clips connected properly, the battery voltage will cause current to flow through diode D1 and relay RL1 coil. If the battery voltage is 6 volts, contacts 3 and 1 remain closed, and current from the positive battery flows through diode D1, relay RL2

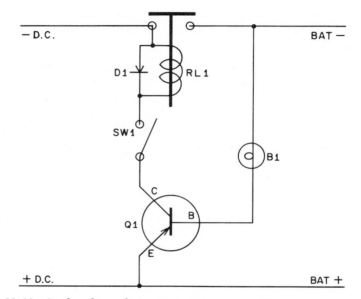

Figure 16–12   Single-voltage alternator protector, source 17

coil (6 volt), through contacts 1 and 3 of relay RL1, and back to the negative battery. Contacts 1 and 2 of relay RL2 close connecting the 115 VAC to the transformer primary, allowing the charger to operate normally.

If the battery voltage is 12 volts, relay RL1 opens contacts 3 and 1, and closes contacts 3 and 2, placing resistance R1 in series with the coil of 6 volt relay RL2 for safe operation on 12 volts. The pilot light B1 is connected to the battery negative from the output side of D1. The pilot light glows when the polarity is right.

If the battery is connected in reverse, diode D1 is blocked and there will be no action to light the pilot light or turn on the 115 volt A.C. Circuit breaker CB1 will operate if the battery has enough power left in it. The clips will spark violently when connected to the battery in reverse, but the 115 VAC will not be connected to the charger, protecting the vehicle alternator diodes for "in the car" charging.

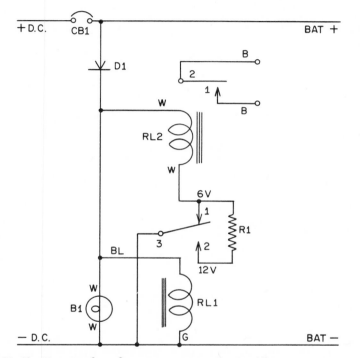

Figure 16–13 Transistorless alternator protector, source 25

The other letters shown on the diagram, not mentioned before, indicate the manufacturer's lead wire colors.

Transistorless Alternator Protector

In Figure 16-14, with the battery clips properly connected and the 115 VAC connected to the charger, the positive battery voltage causes a current to flow through the blue BL lead, ballast bulb B1, diode D1, relay

RL1 coil, and back through the white W lead to the battery negative. This causes relay RL1 contacts to close and allow current flow from the charger positive through black B lead L5, ballast bulb B2, yellow Y lead L6, switch SW1 contacts 3 and 1, through solenoid RL2 coil, red R lead

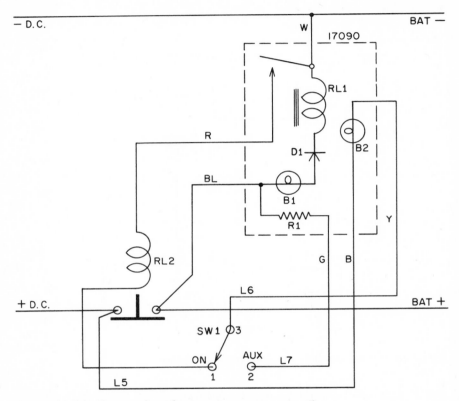

Figure 16–14 Transistorless alternator protector, source 7

through relay RL1 contacts, and back through white W lead to the common negative. Solenoid RL2 contacts close and connect the charger positive to the battery positive, causing normal operation.

With the battery connected in reverse, the diode D1 blocks the flow of current, and nothing happens. Both lights B1 and B2 do not glow. An open circuit in the battery circuit, or a poor connection, would show the same conditions as a reverse connection.

If the battery is dead, or too low to actuate the relay RL1, there is a "dead battery switch" SW1. This SPDT switch SW1 is normally in the "on" position, but has an AUX momentary contact position connecting contact 2 with contact 3. This connects a circuit across the solenoid contacts to give the battery a small charge, so that when SWI is released, the contacts return to the normal "on" position, connecting contacts 3

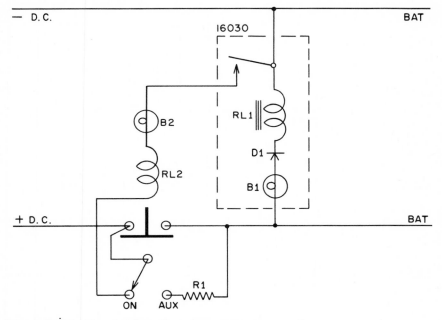

Figure 16–15 Transistorless alternator protector, source 7

and 1. This small charging current circuit goes from the battery charger side of the solenoid RL2 contact, through black B lead L5, bulb B2, yellow Y lead L6, contact 3 and 2 of switch SW1, resistor R1, blue BL lead, and back to the positive battery contact of solenoid RL2.

Transistorless Alternator Protector

Figure 16-15 (part #16030) is the same circuit as shown in Figure 16-14 (part #17090), but it does not have bulb B2 and resistor R1 mounted on the assembly, and they have different part numbers.

Transistorless Alternator Protector

The alternator protector shown in Figure 16-16 connects the 115 VAC to the charger only when the battery polarity is correct, protecting the alternator diodes during "in the car" charging. However, the charger

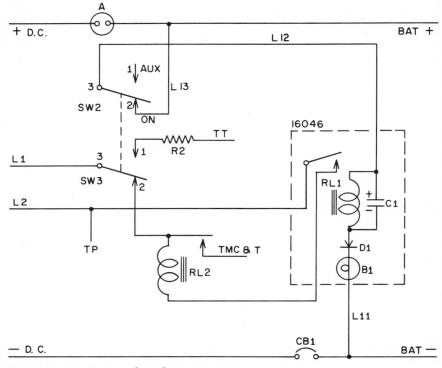

Figure 16–16   Transistorless alternator protector, source 7

diodes must carry the battery reverse current until the circuit breaker CB1 opens, provided that the battery has enough power left in it.

With the battery clamps properly connected and the 115 VAC plugged in, the battery positive voltage causes current flow through lead L13, switch SW2 contacts 2 and 3, through lead L12, relay RL1 coil, diode D1, ballast lamp B1, lead L11, and back to the battery negative. Relay RL1 contacts close and conduct 115 VAC current from line L1, switch SW3 contacts 3 and 2, power relay RL2 coil, relay RL1 contacts, and back to the other AC line L2. Then, power relay RL2 contacts close, connecting the 115 VAC to the charger transformer primary through the timer contacts TMC, and the charger will operate normally.

If the battery polarity is reversed, diode D1 blocks the flow of current, and the 115 VAC is not connected to the charger because relay RL1 contacts are open. However, the battery voltage causes current flow through the charger rectifier in a forward direction, and if the battery has enough power left in it, the circuit breaker CB1 will operate to open the battery circuit.

If the battery is properly connected but is too weak to operate relay RL1, a "dead battery" switch SW2 and SW3 can be operated, ganged together as a DPDT switch. Switch SW2 and SW3, when moved to the momentary contact AUX position, connect contacts 3 and 1 of SW3, opening the circuit to relay RL1, power relay RL2 coil and contacts, and connecting a resistor R2 in series with the line L1 and a transformer primary tap TT to provide a low charging current. When momentary contact switches SW2 and SW3 are released, the charger goes into normal operation, provided that the battery voltage was built up sufficiently during the short AUX charge to actuate the relays.

Alternator Protector

The circuit shown in Figure 16-17 is unique in the way the bias is obtained for base B of transistor Q1, and regulated to be practically constant for different battery voltages.

With the charger clips connected properly to the battery and the 115 VAC cord plugged in, the battery voltage causes current flow from the positive battery terminal through lead L16 to terminal C, to resistor R2, transistor Q1 emitter E, through the parallel circuit of R2 and E to B which is conducting, through series connected diodes D2, D3 and D4, through R3, terminal H, lead 20, and back to the negative battery terminal. Transistor Q1 conducts through emitter E and collector C from the positive battery terminal, terminal C, emitter E, collector C, terminal G, ballast lamp B1 (#1073), terminal F, lead 28, solenoid RL1 coil, switch SW1 contacts 2 and 3, lead 34, and back to the negative of the charger. Solenoid RL1 contacts close and the charger operates normally.

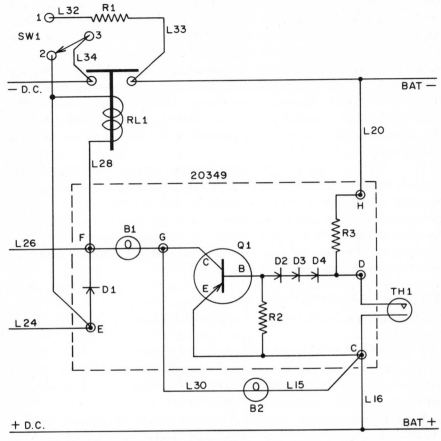

Figure 16–17 Alternator protector, source 7

Pilot light bulb B2 (#1815) is connected across terminal C and terminal G, placing it across transistor Q1 emitter E and collector C. When the battery is connected for correct polarity, the pilot light goes out because of the low voltage drop across the transistor. When the battery polarity is reversed, or there is an open circuit, the pilot light glows. The "dead battery" switch SW1 is SPDT NC on contact 2, and momentarily closed on contact 1. Pressing the dead battery switch SW1 to the AUX position places charging resistor R1 across solenoid RL1 contacts providing a small booster charge to the battery. When the switch is released, the battery voltage should cause the solenoid to close. Diode D1 is a solenoid coil discharge device. Diodes D1, D2, D3, and D4 are identical. When this circuit was designed, it was more economical to use three diodes D2, D3, and D4 in series and operating in the barrier region, instead of a more costly zener diode to control the bias voltage. Thermostat TH1, connected across terminals C and D, is a normally open prod that is inserted in one of the battery cells. When the temperature of the battery reaches about 125 degrees F., the contacts close and short out terminals D and C cutting off the negative bias to the transistor Q1 base B. This brings base B to zero bias by connecting the base B and emitter E through resistor R2. The transistor stops conducting, and solenoid contacts open. Also, it cuts off the 115 VAC through a separate power relay, whose coil is connected in parallel with the solenoid coil, by way of leads L24 and L26. With the polarity reversed, the diodes D2, D3, and D4 block the flow of current and nothing happens except that the pilot light glows.

Transistorless Alternator Protector

The alternator protector shown in Figure 16-18 has a "holding" switch on the solenoid, and a separate pilot light with its own diode.

With the battery clips properly connected, and the 115 VAC plugged in, the positive battery voltage feeds current through lead L5 to the solenoid case, through lead A8 to switch SW2 contacts 3 and 2, lead A9, solenoid RL1 coil, to parallel blocking diodes D1 and D2, ballast bulb B2, and back to the battery negative. Solenoid RL1 contacts close and the charger operates normally. Simultaneously, NO switch SW1, which is part of the solenoid, closes contacts 1 and 2 which are connected across contacts 2 and 3 of switch SW2. So, even if contacts 2 and 3 of switch SW2 are open, the solenoid RL1 and attached switch SW1 will hold the solenoid contacts closed.

Switch SW1 is a steel strap mounted on one of the coil terminals, and bent at a right angle to be within $\frac{1}{16}$ inch of the back center of the solenoid. When the coil of solenoid RL1 is energized, this strap is magnetically attracted to contact the solenoid case. The contact 2 of SW1 is the strap, and contact 1 is the small button on the back of the solenoid case.

The "dead battery" switch SW2 connects an auxiliary charging current across the solenoid contacts. This charging circuit consists of bulb B1 and resistor R1. On later models, in order to increase the auxiliary charging current, a kit #DBK-18 is provided, consisting of leads K-7 and K-6 with resistor R2, that can be added as shown. Diode D3 blocks the flow of current for the pilot light so the light does not glow when polarity is correct.

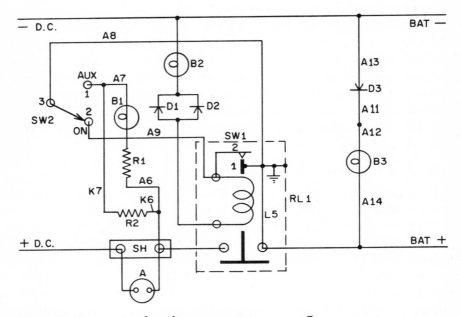

Figure 16–18   Transistorless alternator protector, source 7

With the polarity of the battery clips reversed, the pilot light glows because the diode D3 now conducts and current is fed from the battery positive (now negative) through lead A14, pilot light bulb B3 (a warning buzzer is often used in place of bulb B3), leads A12 and A11, diode D3, lead A13, and back to the battery negative (now positive). Diodes D1 and D2 now block the flow of current, and nothing else happens. Two small diodes D1 and D2 were used in parallel instead of one larger one for uniformity and economy.

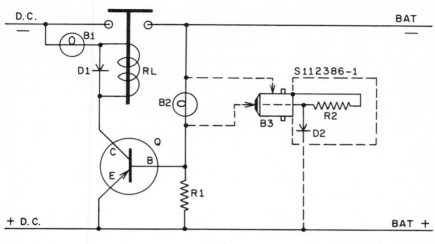

(a)

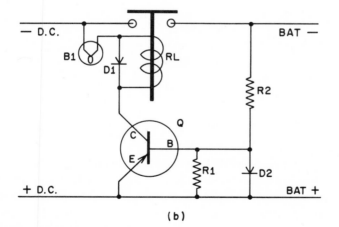

(b)

Figure 16–19   Alternator protectors, source 8

Alternator Protectors

The alternator protector shown in Figure 16-19 is a common circuit, differing only in the use of a #1047 ballast bulb B1, instead of a #1073. The early models, as shown in Figure 16-19a, used a #313 bulb as base lamp B2, which was later changed to the circuit shown in Figure 16-19b. A 50 ohm, 4 watt resistor R2 replaces the #313 bulb, and diode D2 was added to protect the transistor Q on reverse polarity. A replacement for the #313 bulb is available as part #S112386-1, shown in dotted lines in Figure 16-19a, consisting of a miniature bayonet base B3, a 50 ohm resistor R2, and diode D2. The #313 bulb B2 is removed and B3 is inserted in the socket, with the lead going to the battery positive cable or terminal. If the socket is not wired as shown, with the shell going to the battery negative, the wiring must be changed to work properly. This gives the same circuit and values as shown in Figure 16-19b. The following components are the same in both circuits: resistor R1 is 100 ohms, Q is a PNP transistor, RL is a 6 volt, continuous duty, 4-post solenoid, D1 and D2 are small diodes, and resistor R2 is 50 ohm, 4 watt.

17
AUTOMATIC VOLTAGE REGULATORS

Automatic voltage regulators are used in many ways in the charging of batteries. The most common voltage regulator is the one used in vehicles that regulates the output of the generator, or alternator, to prevent over-charging and to supply the load demand as needed. In battery chargers, the voltage regulator is used similarly, but also controls the rate of charge for various conditions, and turns the charger completely off when the battery has reached a set voltage or a maximum fully charged voltage.

Basic operating principles of the various components used in voltage regulators are explained and illustrated in other chapters.

The following diagrams and explanations cover typical automatic voltage regulators used in battery chargers.

Voltage Regulator Using Thyratron Tube

Figure 17-1 shows a typical automatic voltage regulator using a #2050 thyratron VT1 as a voltage control device to maintain constant voltage for a 6 volt battery.

In this circuit, semi-automatic control of a battery charger is obtained by a #2050 thyratron VT1 and a relay RL1 (250 ohms), that shut the charger off at a predetermined (adjustable) battery voltage, and turn it on again when the battery voltage drops to a slightly lower value, maintaining a constant battery voltage for laboratory work, calibration and the like.

Transformer T1 supplies 6.3 volts S1 for the vacuum tube heater H, and also 70 volts for S2, that is fed through relay RL1, and to the plate P of the tube. Relay RL1 turns the charger on when the tube conducts, and turns it off when the tube is not conducting. Battery voltage is fed to the control grid CG, negative, and to the cathode C, positive, through resistors R3 (1K ohms), R5 (1K ohms), and potentiometer R4 (1K ohms). Because, for a 6 volt battery, this bias is too high, an AC voltage is tapped off the 6.3 volts feeding the heater, and placed in series with the battery voltage. These AC voltage peaks reduce the bias across the control

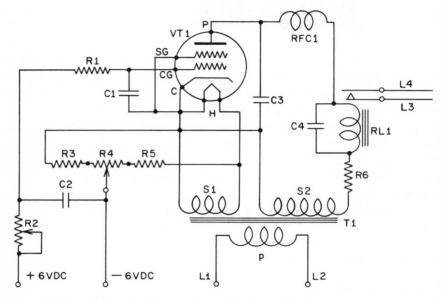

Figure 17–1   Voltage regulator using a thyratron tube

grid and cathode, providing the correct bias to cause the tube to conduct and, at a slightly higher battery voltage, to cut off the charger.

When the control grid CG bias is correct for conduction, the current in the plate circuit is a pulsating D.C. The thyratron rectifies the AC, and conducts during a large portion of the positive peak, but it does not conduct during part of the cycle. Therefore, as long as the bias is correct for conduction, it conducts each cycle, and stops conducting when the bias is too high. Relay coil RL1 does not offer much impedance to the peak current when the coil is by-passed by such a large capacitor C4 (10 mfd). R6 (100 ohms) limits the peak current and, with capacitor C4, prevents hum and chatter in the relay RL1. Condenser C3 (0.1 mfd, 200V) and RFC1 (radio frequency choke) prevent radio frequency radiation.

The grid resistor R1 (0.15 Meg.), R2 (0.5 Meg.), and condenser C1 (0.00025 mfd) prevent slow operation, and false triggering if there is a strong R.F. transmitter nearby, which could cause the relay RL1 to chatter. Capacitor C2 (3 mfd) and resistor R2 form a time-delay circuit to keep the charger from turning off and on rapidly, if the charger is left on a high rate with no load on the battery.

Potentiometer R4 adjusts the bias on the grid CG so it cuts off the charger at the desired battery voltage. The range of 2 to 4 volts across potentiometer R4 is enough to spread the point of conduction in the thyratron from 5 to 7.5 volts on the battery.

Leads L1 and L2 connect to the 115 VAC power. L3 and L4 are the switch points of the relay RL1 and are connected in series with the AC line and the transformer of the charger. The positive 6VDC is connected to the positive 6 volt battery and charger, and the negative 6 VDC is connected to the negative 6 volt battery and charger. The values shown are for a 6 volt charger. However, with slight modification of the values in the control grid bias circuit, this could be adapted to any desired voltage.

Voltage and Temperature Controlled Regulator Using Thyratron Tube

Figure 17-2 shows a combined battery voltage and battery temperature-controlled regulator using 2D21 thyratron tube VT1, called a "Safetronic Control."

The safety therment TH1 is a resistor, encased in a sealed shell and represented by R8 (250K-450K ohms), that changes resistance as the temperature increases, cutting off the charger at a safe temperature. The therment is always placed in the positive cell. The case of the therment provides a negative 2 volts through the white W lead due to contact with the electrolyte in the battery cell. This is the only negative battery connection to the regulator, providing the D.C. voltage to fire the thyratron along with the temperature rise of the battery. The unit works on the same general principles as the regulator shown in Figure 17-1.

The component values are: condenser C1-200 mmfd; C2-0.01 mfd;

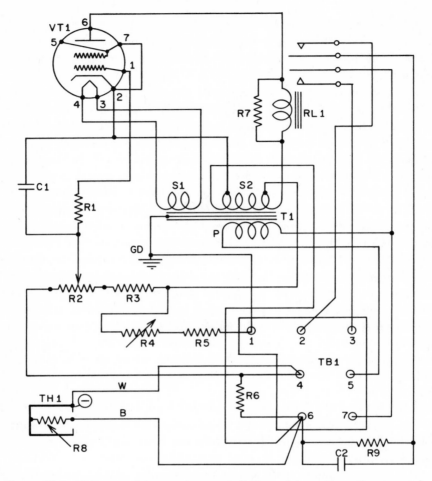

Figure 17–2 Voltage and temperature controlled regulator using a thyratron tube, source 12

resistor R1-100K ohms; R2-3 Meg; R3-1.3 Meg; R4 is a thermistor having a resistance of 82K ohms at 77 degrees F, and 186K ohms at zero degrees F.; resistor R5-200K; R6-1 Meg.; R7-2.2K ohms; R8 represents the internal resistance of the safety therment TH1 and has a resistance of 250K to 450K ohms at room temperature; resistor R9-75K ohms.

Terminal board TB1 is connected to the charger as follows: Terminal 1 to positive D.C. output of charger, T2 to lock-out break on tap or "start" switch, T3 to transformer primary tap, T4 to lock-out break on tap or "start" switch, not occupied by T2, also the white W lead from the safety therment TH1 is connected to T4, T5 to the AC line, T6 to the black B wire of the safety therment TH1, T7 to the "fast" side of the "fast-slow" switch.

The following procedure is recommended to adjust and calibrate the safetronic unit.

A. THE SAFETY THERMENT. This unit is a two wire cable with a temperature sensitive resistor at one end. If either wire is broken, or if the safety therment at the end of the cable is crushed, corroded, cracked, or has any pin holes in it, the complete cable therment should be replaced. At room temperature the resistance of the safety therment should be between 250,000 and 450,000 ohms.

B. THE SAFETRONIC CONTROL. This unit contains a miniature thyratron type 2D-21, a small transformer, a relay which shuts off the fast charge when the battery is fully charged, several resistors and capacitors. The tube may be tested by replacing with a good tube from stock. If trouble isn't corrected with the new tube, and if the safety therment is in good condition, the safetronic control unit should be replaced with a new or factory rebuilt unit.

C. TEST FOR SAFETRONIC CONTROL. Set up charger and rotate charge control switch to start position. Set the "slow-fast" switch to the "fast" position. Proceed as follows:

1. Connect a 115 volt test lamp across AC input terminals on safetronic control terminal board T3 and T5. If the lamp glows, it indicates voltage on the safetronic control unit. If it does not light, the trouble is in the AC cable, toggle switches, the male plug or the receptacle.

2. If the lamp lights, short circuit terminals T3 and T7 with a short wire of about #14 gauge. The charger should operate and the ammeter should indicate a reading normal for the charge control switch position as it is rotated. This is evidence that the safetronic unit is at fault, because the relay contacts are not closing the AC circuit. If the ammeter does not indicate, the safetronic unit is probably all right, and the trouble is elsewhere in the charger.

D. HOW TO CALIBRATE THE SAFETRONIC CONTROL. The safetronic control and its associated therment cable are calibrated at the factory. Replacement of either the control or the therment cable requires recalibration of the entire unit. Equipment required for calibration consists of a mercury bulb thermometer with a range of at least 50 degrees to 150 degrees F. and a charger analyzer.

1. Connect the charger to a good battery, preferably fully charged or nearly so.

2. Follow operating instructions for starting charger. Set slow-fast toggle switch to fast position. Rotate charge control switch to low-medium or high, depending upon the amount of charge required. Use the meter as a guide, but do not exceed 75 amperes.

3. Check and record the ambient (room) temperature.

4. Put the thermometer in the positive cell with the safety therment. If not possible, put the thermometer in an adjacent cell.

5. Maintain the input line voltage at 110 volts AC.

6. Refer to cut-off values Table 17-1, and determine the cut-off temperature for the room temperature previously recorded.

Typical Voltage Regulator Cut-Off Temperatures for Various Room Temperatures

Room Temp. C-O	50	55	60	65	70	75	80	85	90	95	100	105	110
Min. C-O	98	100	103	105	107	109	111	113	116	118	120	122	125
Max.	108	110	113	115	117	119	121	123	126	128	130	132	135

Table 17-1

7. Line voltage affects the cut-off temperature. Add 1 degree to the cut-off temperature for every two volts that the line voltage is above 110 volts and subtract 1 degree for every two volts that the line voltage is below 110 volts. For example, if the line voltage under load is 112 volts, and the room temperature is 70 degrees F., add 1 degree to 107 and 117. Battery charger should then cut off between 108 and 118 degrees. If the line voltage under load is 108 volts, subtract 1 degree from 107 and 117 Battery charger should cut-off between 106 and 116 degrees.

8. A factory-sealed potentiometer is located behind the metal-shielded tube. Break the seal. The potentiometer shaft, when rotated, raises or lowers the cut-off point according to direction of movement. Counterclockwise rotation raises the cut-off point, and clockwise rotation lowers the cut-off point.

9. Rotate the potentiometer shaft fully counter-clockwise.

10. Start the charger for normal charging.

11. As the charging continues, the battery temperature will rise. Watch the thermometer carefully, keeping in mind that you want the charger to cut-off within the temperature range previously specified.

12. When the thermometer reads halfway between the upper and lower limits (halfway between 107 and 117 degrees, or at 112 degrees in our example) rotate the potentiometer shaft slowly clockwise until the charger shuts off. This should be the correct potentiometer setting.

13. This potentiometer setting must be rechecked. Allow the battery to cool slightly. Then, restart the charger. If the setting is correct, the charger should cut-off within a few minutes. If it does not, repeat steps 11 and 12.

14. As a final accurate check, use another battery. Start charger for normal recharge and note cut-off temperature. If the charger cuts off within the ranges previously specified, the potentiometer setting is correct and the safetronic control has been properly recalibrated. Reseal the potentiometer shaft with Glyptol thickened by evaporation.

Voltage Regulator Using a Unijunction Transistor (UJT)

A simple battery charger control circuit using a UJT, is shown in Figure 17-3. A circuit similar to this should appear in future battery chargers because it offers a combination voltage control and protection against reverse polarity and shorts in a single unit with a simple circuit.

The UJT (2N2160), with R1 (3.9K ohms), R2 (2.5K ohms), R3 (3.3K ohms), and C1 (0.25 mfd) form a relaxation oscillator circuit that gets power from the battery being charged, and triggers SCR1 (MCR 808-3) through transformer T1. When the firing voltage of UJT as determined by the battery voltage exceeds the breakdown voltage of zener Z1 (1N573 6.2V, 400 mw), the UJT can no longer oscillate. The charger ceases because SCR1 is not conducting. This circuit cuts off at a specified battery voltage, as determined by the setting of R2, and cannot conduct under conditions of short-circuit, open circuit, or reversed polarity connection to the battery.

The values given for this circuit are for a single voltage, 12 volt 18 ampere charger, that can be modified to operate on any voltage just as effectively.

Voltage Regulator Using Zener Diode and Transistors

A typical voltage regulator circuit shown in Figure 17-4, uses a zener diode Z, a driver transistor Q1, and a power transistor Q2.

The voltage regulator turns the charger off when the battery has been fully charged to a predetermined voltage. It may be shut off completely or designed to keep a full charge by starting the charger again when the battery voltage drops to a lower value.

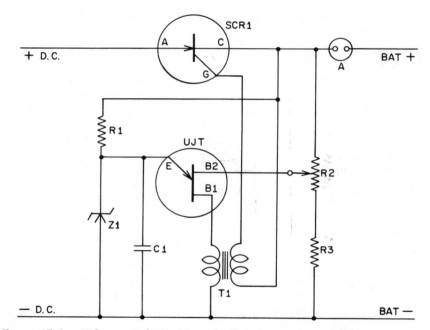

Figure 17–3   Voltage regulator using a unijunction transistor (UJT)

In Figure 17-4, two transistors are used instead of one, permitting the use of a lower current zener Z and a higher current relay, RL1, and resulting in better performance at less cost.

When the battery clips are placed across the battery at the polarity indicated, there is not enough voltage to drive the zener Z into conduction. Therefore, the driver transistor Q1 (PNP) is not conducting. However, the power transistor Q2 (PNP) conducts because the emitter E of the PNP power transistor Q2 is positive and the base B is negative through resistor R4 (820 ohms). This causes the emitter E to collector C to conduct, and the current from the positive line goes through the emitter E to the collector C through the coil of relay RL1 and back to the negative line. This closes relay RL1 contacts RLC1, which are connected in series with the AC line and charger input. The charger comes on and charges the battery to the predetermined battery voltage. By adjusting R2 (600 ohms), the zener diode Z conducts at this voltage.

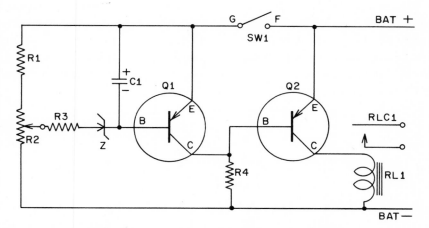

Figure 17–4   Voltage regulator using a zener diode and transistors

The actual voltage to fire the zener is less than the battery voltage, 4.6 volts in this case, due to the voltage divider R1 (910 ohms) and R2 (600 ohms). When the zener conducts, it supplies a negative voltage to the base B of the driver transistor Q1, making it conduct through the emitter E and the collector C. This makes the collector C of Q1 positive because the resistance between the emitter E and the collector C is practically zero, and much less than the resistance of R4 (820 ohms). This makes the base B of Q2 positive and at the same potential as the emitter, since they are directly connected together.

With the base B of Q2 positive, it no longer conducts through E and C and the relay coil, opening the relay RL1 contacts and turning off the charger. It stays off until the voltage drops. Then it automatically starts again. Resistor R3 (1K ohms) and capacitor C1 (10 mfd, 25V) form a filter network to filter out the full-wave rectified ripple and any transient voltages that could prematurely or erratically fire the zener into conduction.

This same basic circuit is used in transistorized voltage regulators on vehicles, except that the alternator or generator field is in the same position as the coil of relay RL1.

Switch SW1, connected between terminals G and F, is part of a timer that connects the voltage regulator only after a preliminary high rate charge from the battery charger.

This voltage regulator is for a single voltage charger. On dual-voltage chargers, such as for 6/12 volts, there is usually a 6/12 volt sensitive relay across the D.C. output that acts as an automatic voltage selector. This relay selects one of two resistor networks in the position of R1 and R2, so that the voltage across the zener diode is the same, regardless of the battery voltage.

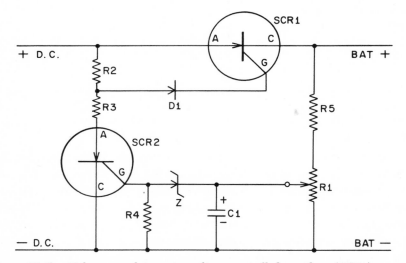

Figure 17–5   Voltage regulator using silicon controlled rectifiers (SCRs)

Voltage Regulator Using Silicon Controlled Rectifiers (SCRs)

Figure 17-5 shows another form of zener controlled voltage regulator that uses two silicon controlled rectifiers instead of transistors; one of them, SCR1, acts as a switch or relay.

When the battery voltage is low, SCR1 receives a positive gate G signal through resistor R2 (27 ohms) and diode D1. SCR1 is therefore conducting during each cycle of the rectified output of the charger, and the battery receives a charging current. When the battery reaches its fully charged voltage and potentiometer R1 (500 ohms) fires the zener diode Z, gate G of SCR2 is made positive, and SCR2 conducts, drawing current through R2 (27 ohms) and R3 (27 ohms). This reduces the gate G voltage of SCR1 so low that SCR1 cannot conduct. Since SCR1 turns off at every half cycle (the only way to turn off an SCR), the gate G voltage is so low it cannot turn on during each cycle. Capacitor C1 acts as a filter to prevent premature, or erratic, firing due to the rectifier output ripple or transient voltages. R5 is 47 ohms and R4 is 1K ohms.

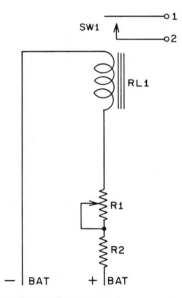

Figure 17–6   Voltage regulator relay, source 4

Voltage Regulator Relay Only Type

Perhaps the simplest voltage regulator circuit is shown in Figure 17-6. It consists of a voltage sensitive relay RL1, similar to those used in vehicle regulators. It is temperature compensated to operate at higher voltages when cold than it does when hot to conform to the battery requirements.

Relay RL1 closes contacts 1 and 2 of relay switch SW1 when the battery voltage reaches the correct full charge voltage. When these NO contacts are closed, they start a second timer in the charger for a finish charge sequence. The cut-off voltage is adjustable by rheostat R1. R2 is a voltage dropping resistor used in some chargers. For example, RL1 and R1 are used on 24 volt chargers, and the same relays RL1 and R1 can be used on 36 volt chargers by adding resistor R2 for an additional 12 volts drop.

Table 17-2 shows typical voltage regulator settings for relay type voltage regulators.

Typical Voltage Settings for Relay Type Voltage Regulators

Regulator Ambient (room) Temperature	24 Volts (12 cells)	36 Volts (18 cells)
125 Degrees F	28.9 Volts	43.6 Volts
115	28.7	43.05
105	29.0	43.5
95	29.3	43.95
85	29.6	44.4
75	29.9	44.65
65	30.2	45.3
55	30.5	45.75
45	30.8	46.2

Table 17-2

Voltage Regulator Using Zener Diode and Transistors

Figure 17-7 shows the basic schematic for all of the terminal voltage relays TVRs shown in Figures 17-8, 17-9, and 17-10.

When the charger is on initial charge, and before the battery reaches the cut-off voltage, the zener diode Z does not conduct, and resistor R5 holds transistor Q1 emitter E at negative potential. At the same time, base B is held at a positive potential through R4 and the bleeder network composed of R6, R7 and R8. Therefore, transistor Q1 cannot conduct through E to C. Relay RL stays in the normally closed NC position, energizing the charger to operate at maximum charge.

As the battery voltage builds up and reaches the trip voltage, the zener diode Z conducts and holds transistor Q1 emitter E at a positive potential.

Base B is held at a negative potential, causing transistor Q1 to conduct from the emitter E to the collector C. This puts a positive bias on transistor Q2 base B, causing Q2 to conduct through C and E to the relay RL coil, opening the NC contacts and closing the NO contacts. The relay RL contacts are held in this position until the battery voltage drops below a lower predetermined value.

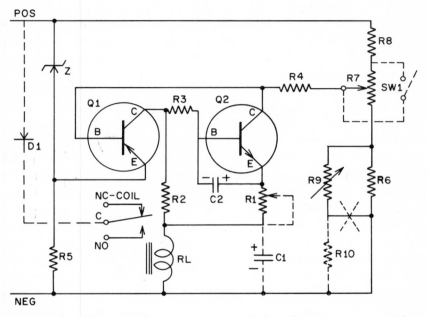

Figure 17–7   Voltage regulator using a zener diode and transistors, source 19

The zener diode Z is a 12 volt, 1 watt unit for all 24 volt regulators, and is a 22 volt, 1 watt unit for all 36 volt regulators.

Potentiometer R7 adjusts the cut-off or trip voltage. Resistor R1 is a fixed resistor on models without a "storage on" provision, but, on other models, it is an adjustable rheostat, shown in dotted lines, that provides "storage on" voltage to restart the charger when the voltage drops to just above the nominal battery voltage.

Switch SW1, shown in dotted lines, is a slide switch, open in the "normal" position, and closed in the "equalize" position. It is used on some models to deactivate the TVR and charge the batteries continuously while equalizing the cells until the switch SW1 is returned to "normal" when the TVR control function takes over.

Condenser C2 is used on all models, but condenser C1 is used only on some models. Resistor R9 is a thermistor that compensates for temperature changes, and on some units, may have a calibrating resistor R10 connected in series. In this case, the connection crossed out with a dotted X would be omitted.

The diode D1 is used on all models having a D.C. contactor. It is wired internally on the TVR in some models, and mounted externally on the D.C. contactor in other models. The diode D1 is not used on models using a 115 VAC contactor.

The zener diode Z, diode D1, transistors, and relay RL coil (1000 ohms) can all be checked in the circuit, using an ohmmeter for the relay coil resistance, and a polarity tester (described in Section II, Chapter 2) for testing the zener diode Z, diode D1, and transistors Q1 and Q2. The zener diode Z zener voltage can also be measured in the circuit. The zener diode is connected in series with a resistor R5 (12K ohms) across the positive and negative terminals. With a high resistance voltmeter across R5, apply an adjustable D.C. voltage across the positive and negative terminals using one of the voltage regulator test sets described in Section II, Chapter 2. Slowly increase the D.C. voltage across the positive and negative terminals until the zener voltage is approached. The D.C. voltmeter across R5 should not start to read until the D.C. voltage reaches the zener voltage, which is 12 volts for all 24 volt chargers, and 22 volts for all 36 volt chargers. When the D.C. voltmeter across R5 just begins to read, the voltage across terminals POS and NEG is the zener voltage. Increase this voltage another 10 volts. The meter across R5 should now read 10 volts, while the meter across POS and NEG should read 10 volts plus the zener voltage. The zener diode is good if it passes this test. If the zener diode is shorted, both meters will read the same at all times. If the zener diode is open, the voltmeter across R5 will read zero at all times when the voltage across the POS and NEG is above the zener voltage.

Potentiometer R7 adjusts the cut-off, or trip, voltage on all models. For typical cut-off voltages, refer to Table 17-1. If there is only one potentiometer R7, the TVR will not turn on automatically after cutting off, unless the voltage decreases to about one half of the cut-off voltage, or the battery and A.C. are both disconnected momentarily. To adjust the single control TVR, turn the potentiometer R7 above the cut-off voltage, or to the extreme clockwise position. Then, turn it down slowly until the cut-off voltage is reached. After it cuts off, disconnect the battery and the AC power to reset it, and check again.

On the dual control type where R1 is adjustable, adjust R1 for the "storage on" voltage (just above the nominal voltage of the battery) to turn the charger back on when the battery voltage is reduced to this value. For a 24 volt battery it should turn on at about 24.6 to 24.76 volts; and on a 36 volt battery, the turn-on voltage should be about 37 to 37.5 volts. This setting requires a very accurate D.C. voltmeter of at least $\frac{1}{2}\%$. If in doubt, discharge the battery slightly to reduce the surface charge to where it reads exactly 24 volts (or 36 volts). This will show, at least, that the turn-on voltage setting is above the nominal voltage of the battery. It is better to have this voltage too high than too low. Otherwise, the battery would be pretty well discharged before it turned on again.

All of the components in dotted lines are used only on some models, while those in solid lines are used on all the TVRs.

The component values are different, in some cases on the same TVR number, from the values given in Figures 17-7, 17-8, 17-9, or 17-10. Replace all defective parts with exactly the same value as the original, as determined by the markings or color codes on the part. Thermistor R9 may be a solid brown color which is a Glo-Bar thermistor having 750 ohms resistance at 25 degrees C., or R9 may be brown with a green dot, which is a Glo-Bar thermistor having 4500 ohms resistance at 25 degrees C. But these markings are not standard and may have other values. If the thermistor is defective, it is best to replace the entire TVR, until more information and parts are readily available for these components. Any deviations, or changes, in the temperature compensating circuit require a temperature controlled chamber to properly recalibrate these units, the cost of which would not be justified. The manufacturer recommends replacing the entire unit in all cases of failure, but this is unnecessary for the qualified technician who is properly equipped with test equipment and parts stock.

Voltage Regulator Using Zener Diode and Transistors

Figure 17-8 shows the physical appearance and lay-out of the parts of terminal voltage relays, number TVR 384 for 24 volts, and TVR 389 for 36 volts. The relay RL contacts C and NC control an AC contactor hav-

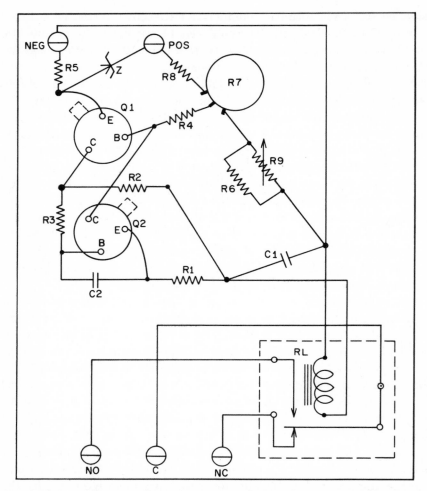

Figure 17–8   Voltage regulator using a zener diode and transistors, source 19

ing a SPDT switch to change from a high charging tap to a lower charging rate tap, and to start the timer in some models.

This is the single adjustment type using potentiometer R7 (600 ohms) to provide only a cut-off trip voltage. The transistors used in early models were Q1-2N464 PNP and Q2-2N213A NPN, while later models use Q1-2N404 PNP, and Q2-2N1302 NPN. The general operation is the same as described for the circuit in Figure 17-7. The values of the other components of one typical model are: R1-1K ohms; R2-1K ohms; R3-330 ohms; R4-2.2K ohms; R5-12K ohms; R6-1.2K ohms; R7-600 ohms; R8-470 ohms; thermistor R9-750 ohms at 25 degrees C.; C1 and C2-10 mfd 50V; RL-1K ohms coil resistance; zener diode Z is 12 volts for 24 volt TVR, and 22 volts for a 36 volt TVR.

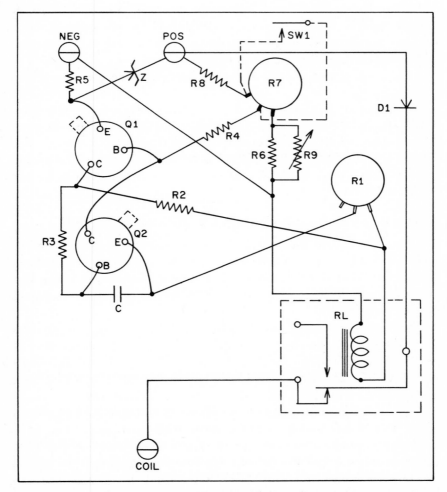

Figure 17–9   Voltage regulator using a zener diode and transistors, source 19

Voltage Regulator Using Zener Diode and Transistors

Figure 17-9 shows the lay-out of terminal voltage relays number TVR 1665 for 24 volts, and TVR 1455 for 36 volts. The relay RL contact terminal COIL controls a D.C. contactor that closes and opens the A.C. and D.C. circuits. It has a built-in diode D1 (1N2069) as a reverse polarity protector.

This unit has a "storage on" adjustment R1 (25K ohms), a "finish off" adjustment R7 (250 ohms), and an "equalize" switch SW1, shown in dotted lines, on some charger models. Switch SW1 is open on "normal", and closed on "equalize" positions.

For operating details, refer to the circuit description given for Figure 17-7.

The values of the components for a typical model are: R1-25K ohms; R2-1K ohms; R3-330 ohms; R4-2.2K ohms; R5-12K ohms; R6-1.2K ohms; R7-250 ohms; R8-680 ohms; thermistor R9-750 ohms at 25 degrees C.; C-10 mfd, 50V; Q1-2N404 PNP; Q2-2N1302 NPN; D1-1N2069; RL-1K ohms coil resistance; zener diode Z is 12 volts for 24 volt TVR, and 22 volts for 36 volt TVR.

Voltage Regulator Using Zener Diode and Transistors

Figure 17-10 shows the lay-out of terminal voltage relays number TVR 3067 for 24 volt, and TVR 3066 for 36 volt. This is similar to that shown in Figure 17-9 except that diode D1 is mounted externally on the D.C. contactor, and all three relay RL contacts are brought out to terminals NO, C, and NC.

The component values are: R1-2.5K ohms; R2-1K ohms; R3-330 ohms; R4-2.2K ohms; R5-12K ohms; R6-1.5K ohms; R7-250 or 600 ohms; R8-1.2K or 470 ohms; thermistor R9-4.5K ohms at 25 degrees C.; C1 and C2-10 mfd, 50V; RL-1K ohms coil resistance; Q1-2N404 PNP; Q2-2N1302 NPN; Z- 12 volts for a 24-volt TVR, and 22 volts for a 36-volt TVR.

For operating details, refer to the circuit description for Figure 17-7.

The TVR shown in Figure 17-10 can be used universally, in an emergency, or to reduce inventory, as a replacement for either of the TVRs shown in Figures 17-8 and 17-9 with slight modifications. To replace the TVR in Figure 17-9 with the TVR in Figure 17-10, add diode D1 between the POS terminal and relay RL contact terminal C, and use

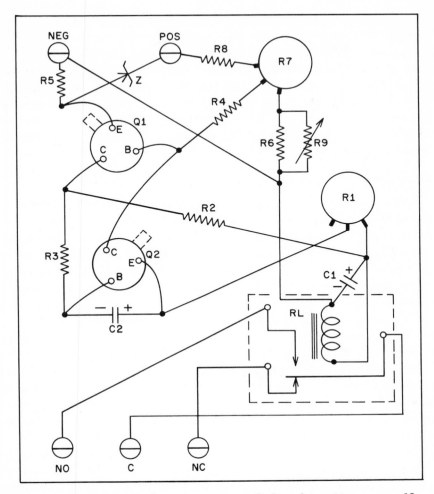

Figure 17–10 Voltage regulator using a zener diode and transistors, source 19

terminal NC for COIL. This diode can be soldered in position, or it can be made up with a quick-connect terminal on each end. Connect one end (cathode) to terminal C, and the other end (anode) to the POS terminal using a 3-way quick-connect adapter.

To replace the TVR in Figure 17-8 with the TVR in Figure 17-10, re-duce the "storage on" adjustment R1 to about one half of the cut-off voltage. The "storage on" adjustment can be used as an added feature. Even though some of the component values may vary between different models, they all will function properly if operated on their designed voltage. Of course, only 24-volt units can be interchanged with other 24-volt units, and only 36-volt units can be interchanged with other 36-volt units. Since the zener diode is the only difference between a 24-volt and a 36-volt unit, they can be interchanged by changing the zener diode, using a 12 volt, 1 watt zener for 24 volt units, and a 22 volt, 1 watt zener for 36-volt units.

Voltage Regulator Using Transistorized Magnetic Saturation

Some voltage regulators, especially those used on large industrial and marine chargers, have a transistorized control to add a controlled D.C. to a saturating transformer, or a saturable reactor, to control the output of the charger. These are called magnetic amplifiers, or modified magnetic amplifiers. They offer a smooth throttling control of the charger output rather than an abrupt reduction such as a relay or contactor would provide.

These units can be designed and calibrated to hold a constant voltage, and automatically adjust to the load requirements on the battery. They also automatically compensate for a wide range of line voltage variations.

A typical unit is shown in Figure 17-11. Transformer T1, diodes D2 and D3, and ammeter A are all conventional. Saturable reactor SR has a primary P2 in series with the power transformer T1 primary P1, and a secondary S2 that is supplied with a controlled D.C. from the rectifier output. With direct current in the secondary S2 of saturable reactor SR, the primary P2 varies its impedance proportionally. Since this primary P2 is in series with the primary P1 of power transformer T1, more or less voltage is applied to transformer T1 primary P1, varying the D.C. output.

A relay RL, having contacts RLC, has its coil connected across the AC power input of L1 and L2. This relay connects the saturable reactor secondary S2 across the D.C. output and battery only when the charger is operating. Otherwise, this secondary S2 would discharge the battery.

With AC and battery properly connected, current will flow from the D.C. positive through rheostat R3, to terminal 3, through SR secondary S2, to terminal 1, and back to the D.C. negative. To vary this current by changes in the battery voltage, an adjustable positive voltage is obtained at the voltage divider R2 and potentiometer R1, fed to terminal 2, through rheostat R4, zener diode Z, to base B of power transistor Q, and out at emitter E, terminal 1, and back to the D.C. negative. If the battery voltage is high enough to reach the zener Z voltage, the transistor passes a current between collector C and emitter E, causing additional voltage drop across rheostat R3. This reduces the current through SR secondary S2, increases

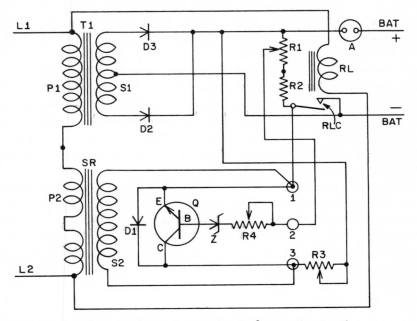

Figure 17–11 Voltage regulator using transistorized magnetic saturation

the impedance of the SR primary P2, and thereby reduces the charger output. Diode D1 is connected with reverse polarity across the transistor collector C and emitter E, protecting the transistor from reverse discharge voltage from SR secondary S2 when the circuit is broken.

The procedure for adjusting this type of automatic control follows. Connect charger to the battery, and set potentiometer R1 for the lowest voltage across terminals 1 and 2 so the zener diode Z will not conduct. Set rheostat R4 in mid position. Connect charger to the AC power. Adjust rheostat R3 for a normal, initial charging current at a nominal battery voltage. If necessary, place a load resistor across the battery to hold the voltage to a nominal 12 volts for a 12 volt battery, for example, while adjusting R3. Remove the load resistor, and allow the charger to build up the battery voltage to 2.3 volts per cell. For a 12 volt battery having 6 cells, this would be a voltage of 13.8. At this point, turn potentiometer R1 slowly to increase the voltage until the zener diode and transistor are conducting, as indicated by a sudden drop in the ammeter A reading. Now, adjust rheostat R4 for the proper trickle charge. This is usually less than 1 ampere, depending on the ampere-hour AH capacity of the battery. This is so low that it will not show on ammeter A. It will be necessary to insert a low range ammeter in series with one lead of the charger and battery. When not actually reading this low range ammeter, it should be shorted out, or removed, to prevent damage.

These adjustments should be rechecked by discharging the battery with a load resistor until the battery is at a nominal voltage on initial charge rate. Also check the initial charging current. Then, remove the load resistor and allow the voltage to rise, and check the trip voltage and trickle charge, making any readjustments necessary.

Without specific information, the value of 2.3 volts per cell should be used for adjusting all voltage regulators that have the charging source permanently connected to the battery to maintain full charge without gassing. Also, this is a good average value to use for vehicle starting batteries that are kept charged by the vehicle alternator, or generator, through a voltage regulator.

18
METERS AND BATTERY TESTERS

The most common electric meters used in battery chargers are "charge rate" indicators (D.C. ammeters) either with or without an external shunt, and "bulb indicators," which are zero center D.C. ammeters with an external shunt. In combination battery chargers and battery testers, there may be an additional calibrated battery testing voltmeter. Also, there may be D.C. voltmeters, A.C. ammeters and A.C. voltmeters, or an ampere-hour meter.

D.C. Ammeters

Three types of ammeters are used on battery chargers: the direct connected D.C. ammeter, the meter mounted shunt type, and the external shunt type.

The direct connected D.C. ammeter, as shown in Figure 18-1a, which has one or more turns of heavy wire in the meter, or the cable on the outside, indicating by induction. The full current goes through the meter to register the amperes. There is no adjustment on this type of ammeter, except a "zero" adjustment in some cases. The dial shows the amperes either in numbers on any of the three types, or by a scale marked in color to show the limits of maximum charge for each battery voltage.

The meter mounted shunt type is shown in Figure 18-1b.

The external shunt type is shown in Figure 18-1c. Both shunt types work on the same principle, and are adjustable by loosening the screw, and moving the connection to change the position of the meter on the shunt. The less metal between the meter terminals, the lower the reading with the same current through the shunt. In both cases, the main charger

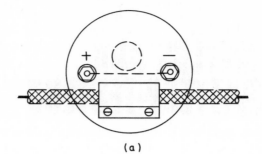

(a)

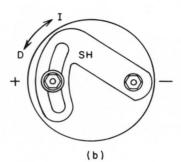

(b)

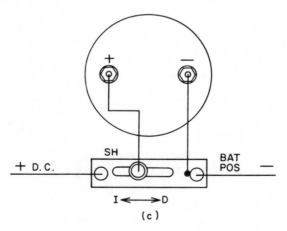

(c)

Figure 18–1   DC ammeters

current goes through the shunt SH. The meter operates from the voltage drop across the shunt. This voltage drop is only a few thousandths of a volt, or millivolts MV. The meter is a millivoltmeter, but it is calibrated in amperes for a particular shunt. The external shunt type avoids running heavy cables to and from the meter. Only a very small wire is needed between the shunt and the meter. The shunt is usually mounted on a heavy stud, such as provided by the circuit breaker, solenoid, or rectifier. In Figures 18-1b and 18-1c, moving the shunt connection in the direction of I is for an increased meter reading, and D for a decreased meter reading.

Voltmeters

Occasionally, D.C. voltmeters connected across the battery charger output are used as charge indicators. When the voltage reaches approximately 13.8 volts for a 12 volt battery, it is considered to be fully charged without gassing at normal room temperatures. Two separate small leads go to the battery charger clips directly, to read the voltage at the battery terminals, avoiding error due to the voltage drop in the charging cable.

Meter Connections

D.C. voltmeters must be connected in parallel to the circuit, that is, positive terminal to positive terminal, and negative terminal to negative terminal. However, an ammeter is series connected so that the positive output of the charger goes to the positive meter terminal. The negative meter terminal would then go to the positive cable clamp. For a negative output charger, the negative output connects to the meter negative terminal, and the meter positive terminal connects to the negative cable clamp.

Usually, the left terminal, while facing the terminals on the back of the meter, is the positive connection on most voltmeters and ammeters.

The snap-on type of plastic meter covers are easy to replace, and often do need replacement. In cleaning plastic meter covers, use only soap or detergent and water. Never use alcohol or cleaning solvent. The cover can be pulled off with the fingers, and snapped back into position. Observe the position of the tabs on the cover and be sure they line up with the proper notches in the meter case. The cover will go on only one way correctly without using force.

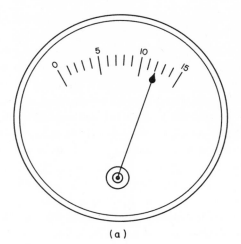

(a)

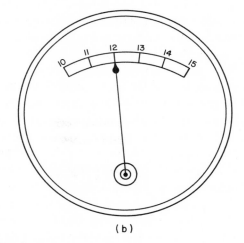

(b)

Figure 18–2   DC voltmeters

Expanded Range D.C. Voltmeters

D.C. voltmeters are used in battery load testers, either as a separate test unit, or built into a combination battery charger and battery tester. The voltmeter leads go directly from voltmeter to the battery charger clips by separate leads to avoid erroneous readings due to the voltage drop in the heavy cables. A battery tester puts a resistance load across the battery being tested, and shows the voltage under load conditions as well as at no load, before and after charging.

The D.C. voltmeter used to show the battery condition and state of charge on most late model testers is a special design known as an "expanded range" or "suppressed scale" D.C. voltmeter.

A standard voltmeter spreads the voltage from zero to full scale over the entire scale. For a 0-15 volt meter for example, the standard linear scale D.C. voltmeter is shown in Figure 18-2a. In battery testing (12 volt), voltages below 10 or above 15 (8-16 are common) are of no particular significance. In Figure 18-2a, the important voltage readings are on the upper third of the scale. The expanded range D.C. voltmeter shown in Figure 18-2b overcomes this disadvantage by expanding the scale 3 times, giving readings 3 times more accurate. The meter in Figure 18-2b does not start reading until the voltage reaches 10 volts.

This expansion may be done in several ways. A bridge circuit using special resistors, or lamp bulbs, is widely used. The bridge circuit is installed either inside the meter, or mounted on a plate external to the meter. In the earliest models of expanded range meters, the suppression was done by using the hair spring adjustment to hold the meter at the low end of the scale until the voltage of about 10 volts was reached. This method resulted in banging the needle against the stop every time the voltage was suddenly removed, often damaging or bending the needle, or causing inaccurate readings. So, these meters were, of necessity, highly damped to check the rebound.

Expanded Range D.C. Voltmeter Using Zener Diodes

The development of the zener diode provided a relatively inexpensive, accurate and simple means to expand the range on a D.C. voltmeter.

In Figure 18-3a, a 10-volt zener diode is shown connected in series with the voltmeter. The meter will get no current until the voltage reaches 10 volts and the zener diode starts conducting, holding a 10-volt voltage drop across it. As the voltage is increased, the zener diode Z has a constant voltage drop of 10 volts, regardless of the current through it, within limits. At a full scale reading of 15 volts, there is a 10-volt drop across the zener diode Z and 5 volts drop across the meter resistance. At a 10-volt reading, there is no voltage across the meter, and a 10-volt drop across the zener diode.

Figure 18-3b shows the same basic circuit, but for dual-voltage opera-

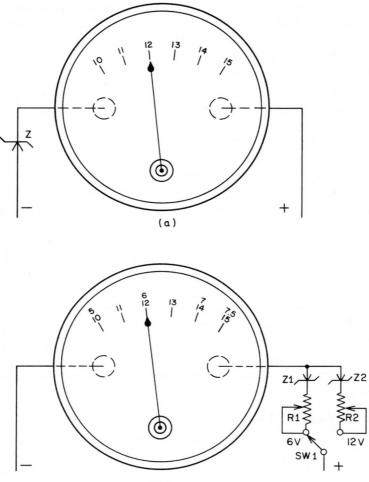

Figure 18–3   Expanded range DC voltmeter using zener diodes

tion, such as 6 and 12 volts, uses two zeners. Zener diode Z1 takes 5 volts or less, and Z2 takes 10 volts or less. R1 is the multiplier and calibrating resistor for 6 volts (5V-7.5V range), and R2 is the multiplier and calibrating resistor for 12 volts (10V-15V range). Switch SW1 is a SPDT switch that selects either the 6 or 12 volt range. Slight deviations of Z1 and Z2 in value can be compensated for by rheostats R1 and R2. Z1 and Z2 should be exactly 5 or 10 volts, or less, but not over. If under, they will start conducting a little sooner.

Sometimes a zener diode will be found connected across the meter terminals. Do not confuse that with the circuits of Figures 18-3a and 18-3b. A zener diode connected across the voltmeter terminals is placed there for "over-voltage" protection. The zener conducts and shorts the meter if the full scale voltage exceeds the safe limits.

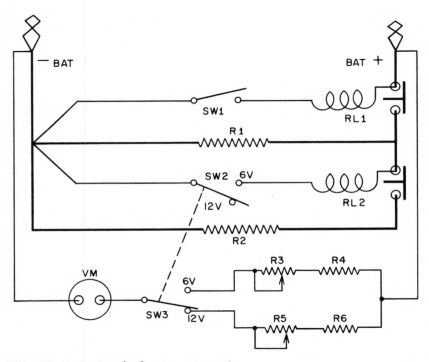

Figure 18–4   Battery load tester, source 12

Battery Load Tester

The battery load tester shown in Figure 18-4 provides a reading of battery voltage under no load, and at a predetermined amperage load on the battery can test 6 or 12 volt batteries. The D.C. voltmeter VM scale is usually calibrated in volts, and always has a colored scale to indicate battery condition and state of charge. It often has a colored scale for calibrating and adjusting voltage regulators on vehicles.

To use this tester, first set the battery voltage test switch, consisting of SW2 and SW3 ganged together as a DPDT switch, to either 6 volts or 12 volts. This selects the correct meter range on the test meter and connects the correct resistance load. When SW1 is set to test, this energizes solenoid RL1, closing the contacts and placing load resistor R1 across the battery. If test switches SW2 and SW3 are in the 12 volt position, solenoid RL2 is not energized, and only resistance R1 is connected across the battery for the 12 volt test. If test switch SW2 and SW3 is in the 6 volt position, solenoid RL2 closes the contacts, and connects load resistance R2 in parallel with load resistor R1 and across the battery. A 6 volt battery needs a higher load test current than a 12 volt battery.

Note that small wires from the test meter circuit go directly to the battery test clips to give accurate voltage readings at the battery terminals. This avoids error due to voltage drop in the cables. Used cables may have many broken strands and offer a higher resistance than a new cable. These small wires are usually built into the heavier cable, or they are taped to the outside of the battery test cables.

Rheostat R3 (25 ohms) is a calibrating resistor, and R4 (39 ohms) is a multiplier for 6 volts. Rheostat R5 (40 ohms) is a calibrating resistor, and R6 (100 ohms) is a multiplier for 12 volts.

Combination Charger and "421" Battery Tester

The battery tester shown in Figure 18-5 is a type "421" tester, and is more accurate and reliable than any previous equipment. It has a built-in 14 ampere battery charger for 12 volt batteries only. This tester has been specifically designed to perform the "421" battery test. This "421" test is the result of extensive testing and analysis of several thousand batteries of all sizes and brands, in various states of charge and condition.

The "421" test is based upon an analysis of differential open circuit terminal voltages. The terminal voltage is measured after the battery has been conditioned by a discharge of 50 amperes for a specific period of time of 15 seconds, and again after the battery has been conditioned by a charge of 14 amperes for a specific period of time of 45 seconds. The difference between these two readings is then compared to a standard

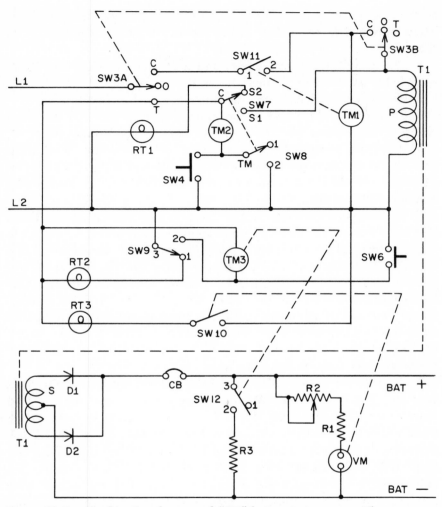

Figure 18–5   Combination charger and "421" battery tester, source 17

colored meter scale to determine battery condition. The standard is based on voltage differences which are characteristic of good batteries and bad batteries. Testers that can do this "421" test can be recognized by the "421 TEST" seal appearing on the instrument panel.

Refer to Figure 18-5 and do the following procedure, which is printed on the front panel of the tester. Explanations have been introduced here, where appropriate, to further explain the use of each component at all stages of the test.

1. Check battery for visual defects, and add water if needed. CAUTION —all lights and accessories must be off.

2. Plug power lead into 115 volt outlet.

3. Connect the red clamp to the positive battery post and the black clamp to the negative post. Twist clamps to make good connection.

The "selector switch", SW3A and SW3B, selects the 421 "test", turns AC "off", or connects the tester as a regular battery charger "charge" at 14 amperes 12 volts. The "selector switch" is a DPDT toggle switch with a middle "off" position.

4. To test, move "selector switch" SW3 to the "test" position. Rotate "test indicator" on the meter (a small knob on the front of the meter) to "up" position. All three test lights, "set", "re-start" and "condition" should be "on". "Test indicator" (SW10) on the meter must be turned to "up" position so that "re-start" light RT3 goes on. Otherwise, the "re-start" light could not give an "off" indication in step 7. Switch SW10 is a micro-switch mounted on the meter that closes turning on "re-start" light RT3 when the "test indicator" is set at over 11 volts on the meter scale. Switch SW10 is open and the "re-start" light RT3 is off when the "test indicator" is set below 11 volts on the meter scale.

5. Depress "set button" SW6, and hold it down until "set light" RT2 goes out. The "set button" SW6 closes the circuit connecting the load timer motor TM3 across the 115 VAC line. After a few seconds, switch SW9, which is a cam-operated switch driven by load timer motor TM3, moves movable contact 3 from contact 1 to contact 2, turning off the "set light" RT2, and connecting contact 3 to contact 2. This shunts the "set button" SW6, and keeps TM3 running, even after the "set button" SW6 is released. At the same time, load timer SW12 moves contact 3 from contact 1 to contact 2, placing the 50 ampere load resistor R3 across the battery for 15 seconds. If the "set button" is not held down until "set light" goes out, the timer and cycle will stop. Press and hold the "set button" until "set light" goes out. After the "set light" goes out, the 50 ampere load is applied for 15 seconds, after which it is turned off The "set light" stays off for 5 seconds more while the battery voltage stabilizes. If the timer stays on and doesn't turn off the "set light" within a few

seconds, immediately disconnect the batteries and check the cam on the load timer TM3. The load timer can "hang up" on the cam front and stall the timer motor leaving the load resistor connected across the battery. Usually, a small dab of grease on the steep part of the cam will cure this problem. If not, smooth off the cam to remove any burrs. The recommended grease is Molykote G lubricant. Run the tester through the load test cycle, *with the batteries disconnected,* as given in step 5, to be sure the timer is operating properly several times. Better yet, disconnect the timer motor TM3 leads and apply 88 volts through a Variac, potentiometer, or other adjustable A.C. source. If it operates for several minutes on 88 volts, it can be considered to be dependable on 115VAC. The torque at 88 volts is only 58½% of the torque at 115 volts. The torque varies, not proportionally, but with the square of the voltage.

The large contacts 2 and 3 of load timer switch SW12 and the load timer micro-switch SW9 are driven by load timer motor TM3 by two separate cams. After 20 seconds, timer TM3 returns the contacts of SW9 and SW12 to the original position as shown in Figure 18-5.

If the load timer should be left "hung up" with the load resistor R3 still connected, do not try to charge the battery or to do step 7, because the extra load on the rectifiers would overload them or kick out the 30 ampere circuit breaker CB.

6. When "set light" comes on again, immediately rotate "test indicator" until it crosses the meter pointer at the "set line".

7. If "restart" light goes off, proceed to step 9. The "restart" light will not go off if setting of "test indicator" in step 6 was over 11 volts.

8. if "re-start" light stays on, repeat steps 5, 6, and 7. If "re-start" light stays on after repeating three times, proceed to step 9.

9. Depress "condition button" and hold it down until "condition" light goes off. "Condition button" SW4 operates condition timer TM2, starting a 14 ampere, 45 second charge cycle. Charge starts when "condition" light RT1 goes off and ends when "condition" light glows. When "condition button" SW4 starts the condition timer motor TM2 after a few seconds, the timer drive cam switch SW7 moves contact C from contact S2 to contact S1. This disconnects "condition" light RT1 and it goes out. Contacts C and S1 apply 115 VAC to the primary of the transformer T1, and puts a 14 ampere charge on the battery through silicon diodes D1 and D2 in a conventional center-tapped, full-wave charger circuit. At the

same time, the condition timer motor operates switch SW8 by moving contact TM from contact 1 to contact 2, shunting "condition button" switch SW4. Motor TM2 will continue to run through the cycle, even after "condition button" switch SW4 is released. SW4 can be released as soon as, but not before, "condition" light RT1 goes out. The condition timer motor runs for 45 seconds and then it shuts off, stopping the charging cycle by moving contact C away from S1 half-way between S1 and S2. The condition timer motor continues to run for an additional 15 seconds to stabilize the voltage. Then, SW8 shifts movable contact TM from contact 2 to contact 1, stopping the motor TM when movable contact C of SW7 moves to contact S2, turning on the "condition" light RT1.

10. When "condition" light glows again (in about 1 minute), immediately note where the meter pointer crosses "test indicator" line. Reading in left red area indicates low voltage after charge due to a possible short in the battery.

Reading in right red area indicates high voltage after charge due to sulphation or high battery resistance. Reading in the green area indicates a good battery. If recharging is necessary, the approximate charge time in hours can be read.

11. If in either red area replace battery. If in the green area battery is "OK"; recharge number of hours shown on upper scale.

BATTERY CHARGE PROCEDURE. 1. With the selector switch in "off" position, connect cable clamps to battery with correct polarity. Move selector switch to "test" position and all three lights should glow before switching to charge. Move the selector switch to "charge" position.

2. Set "charge timer" TM1 to desired number of hours.

In Figure 18-5, the 12 hour timer switch SW11 is a cam-operated switch driven by "charge timer" motor TM1. When the charge timer dial is turned clockwise, it closes contacts 1 and 2, connects charge timer motor TM1 and transformer T1 primary across the 115 VAC line, and continues to run the number of hours set on the dial. This provides a 14 ampere (at start), 12 volt charge to the battery. The test voltmeter VM can be calibrated to read the exact voltage across the cable clamps, with the "selector switch" at the "off" position, by adjusting rheostat R2 which is located inside the case and has a screwdriver slot in a stub shaft. Of course, a precision $\frac{1}{2}$% voltmeter should be used to calibrate the meter VM. However, fairly accurate results can be obtained by using a fully charged battery that has stabilized for 24 hours after charge without an additional charge or discharge during the 24 hour period. At 77 degrees F., the battery should have a voltage of 2.1 volts per cell, or 12.6 volts for a 12 volt battery. If the meter VM on the tester is fairly close, leave it as it is. If the test meter is far off, adjust rheostat R2 to read exactly 12.6 volts on meter VM.

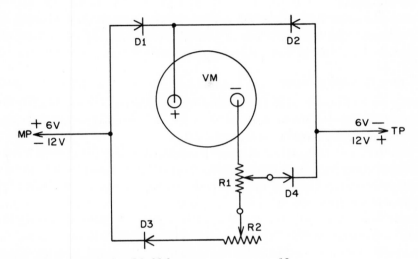

Figure 18–6   Simple hand-held battery tester, source 12

Simple Hand-Held Battery Tester

Figure 18-6 shows a small hand-held voltmeter used to test battery voltage under starter load, as well as for setting voltage regulators on vehicles. This tester is unusual because it does not use a switch to read 6 or 12 volts. Just reverse the connections. There is a meter prod MP protruding from the case of the meter and a test prod TP at the end of a flexible cable. To read on the 6 volt scale, connect the meter prod MP to the battery positive post and the test prod TP to the battery negative post. On the 6 volt scale, the current flows through the meter prod MP and silicon diode D1, but is blocked by diode D3 and D2, through the positive meter VM terminal, through 6 volt multiplier potentiometer R1, and through diode D4 to the test prod TP, and back to the battery negative post. To read on the 12 volt scale, reverse the prod connections used above. The test prod TP is connected to the positive battery post and the meter prod MP is connected to the negative battery post. On the 12 volt scale, the current flows through the test prod TP, through diode D2 (D4 and D1 are blocking), through the positive meter terminal through R1, R2 and D3 back to the negative battery post through meter prod MP.

Diodes D1, D2, D3, and D4 are all silicon diodes rated at 500 MA, 50 PIV or higher.

Potentiometer R1 is the 6 volt calibrating resistor (50 ohms). It should be adjusted first in case there is a small reverse leakage in diode D4. Rheostat R2 is the 12 volt calibrating resistor (100 ohms).

Battery Condition Tester

Figure 18-7 shows a battery condition tester having a timed load of 250 amperes at 6 volts, 180 amperes at 12 volts, an expanded scale meter, and an automatic 6-12 volt selector relay RL3.

The 6-12 volt relay RL3 has a set of double contacts 1 and 2, and a set of triple contacts 3, 4, and 5. With the coil of RL3 de-energized, or on 6 volts, the contacts remain in the 6 volt position as shown. However, 12 volts will energize the armature and move contact 1 away from contact 2, and contact 3 away from contact 4 to contact 5, thereby setting the load and meter to the 12 volt range. Solenoid RL1 closes during the "battery condition" or load test for both 6 and 12 volts, through either the timer TM contacts SW1 contacts 1 and 2, or by "battery condition" switch SW3A contacts 1 and 2, or both. This connects the 180 ampere discharge resistor R1 across the battery. For 6 volts, switch RLC3 connects contact 1 and 2, energizing solenoid RL2 and placing discharge resistor R2 in parallel with R1 and across the battery for the higher 250 ampere load on 6 volts.

The "state of charge" and "battery condition" switch SW3A and SW3B is a DPDT switch. When in the "state of charge" position, contacts 1 and 3 of SW3B are closed placing the voltmeter VM across the line for "state of charge" and "open circuit" readings. The "battery condition" position closes contacts 1 and 2 of SW3A, energizing the discharge solenoids, and contacts 1 and 2 of SW3B energizing the test meter VM through a lower resistance multiplier network so that the condition voltage will rise higher

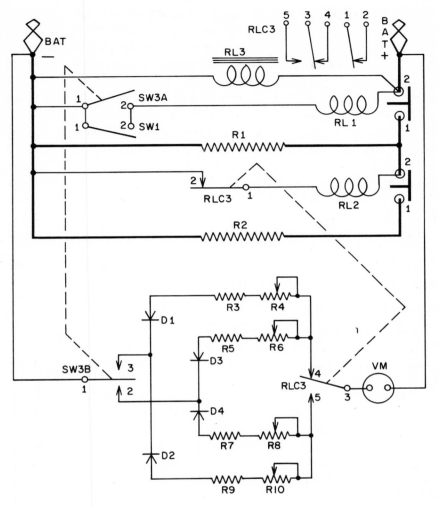

Figure 18–7   Battery condition tester, source 12

on the scale. This is an expanded range meter. The four silicon diodes D1, D2, D3, and D4 are similar and are used to simplify the mechanical switching of the voltage multiplier circuits of the voltmeter. They are automatic switches. For example, on the 6 volt open circuit and "state of charge" position, the current flows from the positive cable clamp through the meter, switch RLC3 contacts 3 and 4, through R4 (25 ohms), R3 (33 ohms), diode D1, through contacts 3 and 1 of SW3B, and back to negative (D2 blocks current). On the 6 volt "battery condition" position, the current flows from the positive cable clamp through the meter, RLC3 contacts 3 and 4, through R6 (25 ohms), R5 (18 ohms), D3, SW3B contacts 1 and 2, and back to the negative cable clamp. On the 12 volt "state of charge" or open circuit position, the current flows from the positive battery clamp, through the meter, through RLC3 contacts 3 and 5, through R10 (40 ohms), R9 (100 ohms), D2, switch SW3B contacts 3 and 1, and back to the negative battery clamp. On the 12 volt "battery condition" position, the current flows from the positive battery clamp, through the meter, RLC3 contacts 3 and 5, R8 (40 ohms), R7 (75 ohms) D4, SW3B contacts 2 and 1, and back to the negative battery clamp.

The test procedure, and interpretations of the readings, are printed on the front of the panel.

Solenoids RL1 and RL2 are 6 volt intermittent duty units, but will stand 12 volts for the few seconds they are energized.

Battery Overall Tester

Figure 18-8 shows the schematic of a battery universal tester. This universal tester can do a number of different battery tests.

The various test procedures found printed on the front panel of the tester, are as follows:

TEST #1 BATTERY CHARGE. Set "car battery" switch SW1, A, B, C, and D, to type of car (6V, 12 volt compact or 12 volt standard). When contact 4 of each of the switches SW1, A, B, C, and D is set on contact 1, the circuit is set up for 6 volt batteries; when set on contact 2, the circuit is set up for 12 volt compact car batteries; and when set on contact 3, the circuit is set up for 12 volt higher capacity standard car batteries.

Connect red clamp to positive battery post, and black clamp to negative battery post.

Depress load switch SW2, A and B for 15 seconds to remove the surface charge. If switch SW1, A and B are in the #1 position for 6 volts, solenoid RL1 and RL2 are energized, and discharge resistors R1 and R2 are connected across the battery in parallel at 165 amperes load. With the switch SW1, A, and B in #2 position, only solenoid RL1 is energized connecting load resistor R1 across the battery placing a load of 150

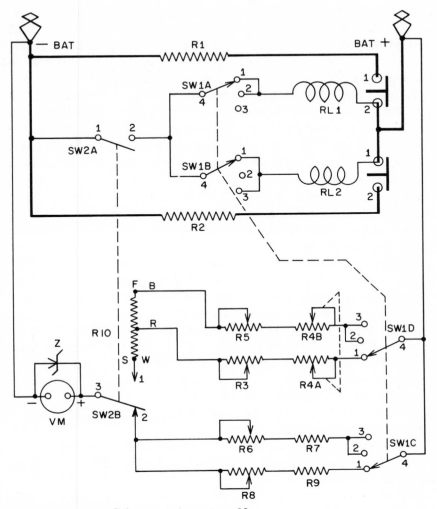

Figure 18–8 Overall battery tester, source 12

amperes on 12 volt compact car batteries. With the switch SW1, A and B in the #3 position, only solenoid RL2 is energized, placing discharge resistor R2 across the battery at 180 amperes for standard size 12 volt car batteries. Meanwhile, switch SW1, C and D connects the test meter through series resistors across the battery.

In position #1 (6V), SW1D connects the positive clamp through contact 4 of SW1D and contact 1, through R4A (32 ohms), R3 (8 ohms), R10 (see Table 18-2) red to white, through contacts 1 and 3 of SW2B, through the test meter, and back to the negative clamp. A zener diode Z is placed across the test meter to protect against reverse polarity and transient voltages above 16 volts. R10 is a temperature compensating probe, consisting of a sealed spool of nickel wire whose resistance changes with the temperature of the electrolyte in the battery. This probe must be placed in any cell, during the load test only. This probe changes the calibration of the load scale, depending on its temperature. It is not in the circuit during the open circuit test. Refer to Table 18-2 for temperature-resistance values.

On positions #2 and #3, switch SW1D connects the positive clamp to contact 4 with 2 and 3, through R4B (90 ohms), R5 (25 ohms), R10 black to white, through SW2B contacts 1 and 3, test meter, and back to the negative clamp.

Depress load switch for 15 seconds to remove surface charge. While battery voltage is settling, then, remove vent caps and check water level.

Read center "battery charge" scale. For a red area reading, recharge battery. For a green area reading, charge O.K.

TEST #2 BATTERY CONDITION. NOTE: Do this test only if test #1 indicates 25% or more charge in battery. If less than 25% charge, recharge battery, then perform #2 test.

Insert temperature compensating probe R10 into any cell of the battery to compensate for temperature of electrolyte.

Connect red clamp to positive battery terminal and black clamp to negative battery terminal.

Set car battery test switch SW1, A, B, C, and D as in test #1.

Set "percent charge" switch (indicator) to reading (25% or more) obtained in test #1.

Depress "load switch" and read lower center "battery condition" scale while depressing "load switch".

Set "percent charge" switch (indicator) to "after charge" arrow, if battery has been fully charged. If red area reading, replace battery. If green area reading, battery is O.K.

TEST #3 CHARGING SYSTEM UPPER RIGHT SCALE. For all fully charged batteries, set battery switch SW1 as in test #1 and #2. Connect red clamp to positive, and black clamp to negative battery posts. Turn all electrical accessories off. Warm up engine, and operate at fast idle for 5 minutes. Then, with car still at fast idle, read upper right hand scale.

Red High indicates that the voltage regulator needs adjusting to prevent overcharging and boiling, or other damage to the battery.

Green OK indicates that the alternator, or generator, and voltage regulator are O.K.

Red Low indicates that the fan belt is slipping, or the alternator, or generator, or regulator are faulty.

Check the fan belt first. Check generators by disconnecting the field (F) lead at the regulator, and touching the disconnected wire, first to ground, then to the terminal on the regulator marked "GEN" or "ARM".

Check alternators by disconnecting the field (F) lead at the regulator, and touching the disconnected wire to the "IGN" or "SW" terminal on the regulator.

If the above generator/alternator tests result in a Green OK or Red High reading, replace the regulator. If not, replace the generator or alternator.

To check or calibrate the meter, refer to Figure 18-9b for the correct voltage, with tolerances, at check points shown in Figure 18-9a.

TEST #4 STARTER. Battery must be fully charged.

Set "battery switch" for correct battery voltage (6V or 12V).

Connect red clamp to positive, and black clamp to negative battery posts. Turn all electrical accessories off.

Crank engine with remote starter button, leaving ignition key off. Engine may be cranked with the key or car starter button, but the center wire should be removed from the ignition coil to prevent the engine from starting.

Red Low indicates bad bearings or defective armature in starter motor.

Green indicates that starter is OK.

Red High indicates worn brushes, loose connections, or bad cables.

TEST #5 SHORT OR LEAKAGE IN CAR'S ELECTRICAL SYSTEM. Turn off all accessories.

Remove ground cable from battery. Set battery switch for correct battery voltage (6V or 12V).

For cars with a negative ground, connect black test clamp to the negative battery post, and the red clamp to the ground cable or to the engine block.

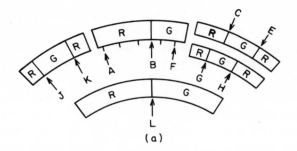

(a)

POINT	METER CALIBRATION VOLTS	INSTRUMENT CALIBRATION	
		6 VOLT SCALE	12 VOLT SCALE
A	1.95 ± .02	5.85 ± .06	11.70 ± .12
B	2.05 ± .02	6.15 ± .06	12.3 ± .12
C	2.3 ± .05	6.9 ± .15	13.8 ± .30
E	2.52	7.55 ± .30	15.1 ± .60
F	2.11	6.33 ± .06	12.66 ± .12
G	2.25	--------	13.5 ± .30
H	2.47	--------	14.82 ± .60
J	1.48	4.44 ± .6	8.88 ± 1.2
K	1.83	5.49 ± .15	10.98 ± .30
L*	1.74/1.60	5.22 ± .06	9.60 ± .12

* AT 75 DEGREES F., AND WITH COMPENSATOR SET AT 75%;
SEE TABLE 18-1 FOR OTHER TEMPERATURES.

(b)

Figure 18–9   Dial and calibration chart for overall battery tester (Figure 18-8), source 12 Model 260N

For cars with positive ground, reverse above connections; that is, connect the red test clamp to positive battery post, and the black clamp to ground cable, or engine block.

Any reading at all indicates a drain on the battery. Touch ground cable to its battery post to wind the electric clock, if any.

Calibration—Model 260N, Load Circuit Only

| Temperature Degrees F. | VOLTS D.C. FOR "L" SCALE REFERENCE | |
	12 Volts (0.0138 V/°F)	6 Volts (0.0057 V/°F)
62	9.42	5.15
64	9.45	5.16
66	9.48	5.17
68	9.50	5.18
70	9.53	5.19
72	9.56	5.20
74	9.59	5.21
75	9.60	5.22
76	9.61	5.22
78	9.64	5.23
80	9.67	5.24
82	9.70	5.25
84	9.72	5.26
86	9.75	5.27
88	9.78	5.28
90	9.81	5.29
92	9.83	5.30
94	9.86	5.31
96	9.89	5.32
98	9.92	5.33
100	9.94	5.34

Table 18-1

Then, if the meter continues to read, there is a short or leak in the car's electrical system.

When checking systems for shorts or leakage, leave the meter connected to indicate when fault is located.

In tests 3, 4, and 5, the test meter is connected and used on the O.C. (open circuit) connection, and meter multiplier resistors R6 (40 ohms), R7 (100 ohms), R8 (25 ohms), and R9 (39 ohms) are in the circuit. When the load switch is depressed, the voltage drops. To raise the meter pointer on the scale, lower multiplier resistors R3 (8 ohms), R4A (32 ohms), R4B (90 ohms), R5 (25 ohms), and R10 (only used on load test) are used. Resistance values for R10 are shown in Table 18-2 for various temperatures.

Temperature Compensating Resistor R10 Fig. 18-8
Model 260N

Temperature Degrees F.	12 Volt Comp. Res. Ohms (white to black)	6 Volt Comp. Res. Ohms (white to red)
62	53.53	23.1
64	53.82	23.1
66	54.11	23.3
68	54.50	23.4
70	54.79	23.5
72	55.08	23.7
74	55.37	23.85
75	55.515	23.9
76	55.66	23.9
78	55.95	24.00
80	56.24	24.20
82	56.53	24.35
84	56.82	24.40
86	57.11	24.60
88	57.40	24.65
90	57.69	24.80
92	57.98	24.90
94	58.27	25.00
96	58.56	25.20
98	58.85	25.30
100	59.14	25.40

Table 18-2

Figure 18-9a shows the expanded range meter scale, and Figure 18-9b shows the calibration voltages for the test points indicated. Table 18-1 shows the calibration of the "L" scale for various temperatures, and Table 18-2 shows the resistance values of R10 for various temperatures. All of the above figures and tables apply to schematic of Figure 18-8.

BATTERY LOAD AND CELL TESTER. Figure 18-10 shows a battery tester having two prods PR1 and PR2 to read individual cell voltages on all but hard-top batteries.

The prods measure the voltage of one cell at a time, or on the car

starter load test, or open circuit, or charging test. All cells should read
the same voltage. A low, or high, reading indicates a bad cell. Nominal
open circuit (unloaded) voltage should be about 2-2.1 volts per cell.
Do not attempt to use the prods while the large clamps are connected

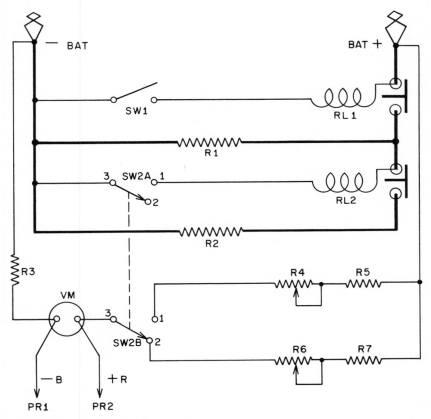

Figure 18–10 Battery load and cell tester, source 12

to the battery terminals. The bottom scale indicates the cell voltage
from 1.4 to 2.6 volts. There is no calibration for the meter alone, but
this is not important because the main concern is the comparison between
cell voltages, and not the exact voltage. A calibration is provided for
the 6 and 12 volt ranges, however.

The procedure to use is as follows:

Set battery selector switch SW2A and SW2B to voltage of battery, 6 or 12 volts.

1. Close timer switch SW1, which runs for 15 seconds and shuts off. When SW1 closes, it energizes solenoid RL1 and places discharge resistor R1 across the battery for either 6 or 12 volts. If switch SW2A and SW2B are in the 6 volt position, solenoid RL2 is energized also, placing both R1 and R2 across the battery in parallel. The voltmeter VM reads the voltage across the battery. When SW2 is in the 6 volt position, the multiplier resistors R4 (25 ohms) and R5 (9.1 ohms) are in the meter circuit. When SW2 is in the 12 volt position, resistors R6 (40 ohms) and R7 (75 ohms) are in the meter circuit.

2. Read upper center scale after timer shuts off.

AFTER CHARGING BATTERY.

1. Close timer switch SW1.

2. After timer shuts off, close timer switch again.

3. Read left scale while timer is on.

GENERATOR REGULATOR TEST.

1. Turn all lights and accessories off.

2. Read battery voltage in upper right scale.

The meter can be calibrated at the line between LOW and SAFE under the generator-regulator test (about 6.75 volts on the 6 volt position and 13.5 volts on the 12 volt position). At the line between SAFE and HIGH, the voltage should be 7.1 and 14.2 respectively. Potentiometer R4 (25 ohms) is the calibrating potentiometer for 6 volts, and R6 (40 ohms) is the 12 volt calibrating potentiometer. Resistor R3 (30 ohms) prevents an accidental short if the clips are left on the battery during the cell test.

Battery Condition Tester

The battery tester shown in Figure 18-11 has a voltage sensitive relay RL3, as an automatic battery voltage selector to hold the contacts NC as shown for 6 volts. But 12 volt batteries operate relay RL3 to open the NC contacts 2-3 and 4-5 and close the NO contacts 1-2. On 6 volts, the solenoid RL1 closes, placing load resistor R1 across the battery when discharge switch SW1 is pushed, and the meter VM reads the battery condition. Meter multiplier resistor R4 is short circuited by NC contacts 4 and 5 on 6 volts, but is in the circuit for 12 volts.

On 12 volts, the relay RL3 opens contacts 4-5 and 2-3, and closes contacts 1 and 2, so that when SW1 is pushed, the 12 volt solenoid RL2 closes and places load resistors R1 and R2 in series across the battery. With the discharge switch SW1 open, the voltmeter VM reads the voltage of the battery under no load conditions.

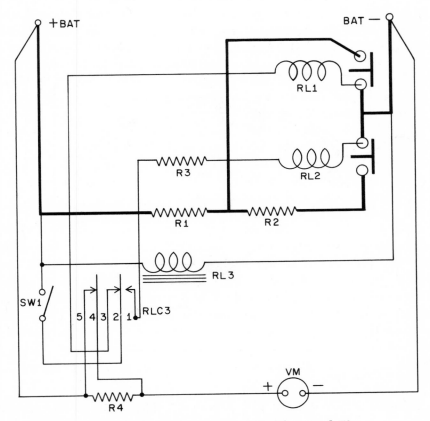

Figure 18–11 Battery condition tester, source 8 Models T2 and T3

Meter calibration is shown in Figure 18-12 for models T2 and T3, and in Figure 18-13 for late model T3 tester. There is no calibrating potentiometer for the voltmeter, but it can be adjusted by turning the screw protruding from the face of the meter cover. If there is no adjusting screw, remove the plastic cover and adjust the hair spring HS tension by moving adjusting arm AA, which extends from the pivot point and is connected to the hair spring. It should not be necessary to loosen the pivot nut PN to make this adjustment.

A ½% calibrating D.C. voltmeter is recommended to make this adjustment, but if one is not available, the following procedure should be followed:

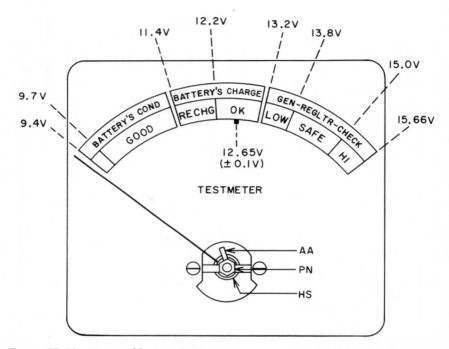

Figure 18–12 Meter calibration for battery tester, source 8 Models T2 and T3

Connect a fully charged 12 volt battery having 1.250 specific gravity or more at room temperature that has not been charged or discharged for at least 24 hours. The test meter should be adjusted to read in the middle area of "OK" under "Battery's Charge". This should be 12.65 volts, plus or minus 0.1 volt. More precisely, the voltage should be 12.2 volts, if using a calibrating meter, at the line between "Recharge" or

"LO", and "OK" under "Battery's Charge". Other calibrating points are shown with the correct voltage V shown.

These are expanded range D.C. voltmeters. Therefore, do not attempt to set the pointer at the far left of the scale as in a zeroed type meter, but set it at the voltage indicated.

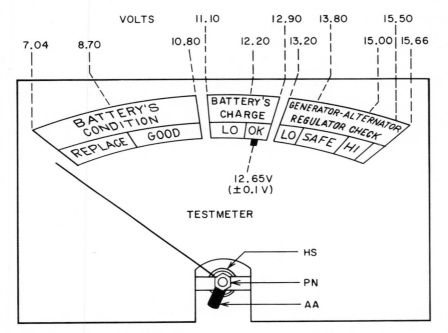

Figure 18–13   Meter calibration for battery tester, source 8 new Model T3

Meterless Battery Cell Comparison Tester Model T5

Meterless battery testers are more portable, less fragile, and less expensive than those using calibrated meters. These testers operate by reading a dial calibrated potentiometer. They use SCRs or zener diodes and transistors, and usually are temperature compensated by a thermistor. A typical model T5 meterless tester is shown in Figure 18-14. Complete instructions for its use are printed on the panel. The calibration can be checked and recalibrated by adjustment controls provided. Number 1 on the dial represents 2.15 volts and number 21 represents 1.95 volts,

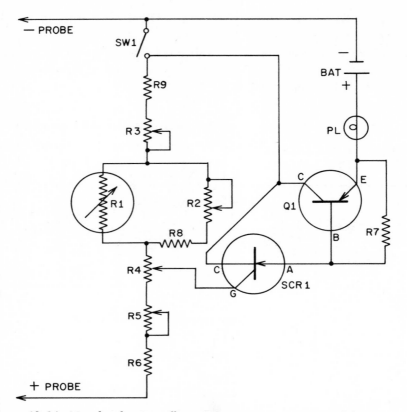

Figure 18–14 Meterless battery cell comparison tester, source 8 Model T5

with each numbered division in between representing 0.01 ($\frac{1}{100}$) volt. Number 11 on the dial is 2.05 volts, and is also the dividing line between "Charge OK" and "Needs Recharging". The prods are placed in two adjacent cells, one at a time, and the readings compared. The dial is turned to number 1 and the button is pressed. Then, the dial is slowly turned CW until SCR1 fires and lights the pilot light PL. Resistor R1 is a temperature compensating thermistor. The battery BAT is a 1-½ volt size AA dry cell, installed inside the tester. It should be re- placed when the light gets dim.

To replace the battery BAT or to make internal repairs or adjustment, open the case which is made in two parts, and locked together in the four bottom corners with nylon pressure studs which must be removed.

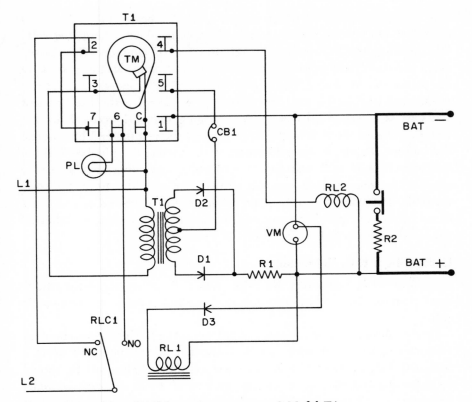

Figure 18–15 Type "421" battery tester, source 8 Model T4

Remove them by inserting a screwdriver between the two parts of the stud head, and prying upward.

"421" Battery Tester

All "421" battery testers perform the same functions and cycles, but various manufacturers have different ways of doing the same thing.

The Model T4 "421" battery tester is shown schematically in Figure 18-15. Complete operating instructions are on the panel, and the timer TI dial is clearly marked as to the cycle. The pilot light PL burns at three reading points. If it lights at any other time or when first hooked up, the tester is connected with reverse polarity. The relay RL1 coil is connected in series with diode D3 so that with correct polarity the relay RL1 does not operate. Diode D3 blocks current so the RL1 contacts stay in the NC position, which connects the AC power to the timer TI contact 2. On reverse polarity, diode D3 now conducts, energizing relay RL1, opening the NC contacts and closing the NO contacts. This disconnects the timer and power to the charger, and connects the pilot light PL across the AC line.

Solenoid RL2 is the load test contactor to energize the load resistor R2; CB1 is a circuit breaker in the charging line; D1 and D2 are the charging diodes through transformer T1; and the test meter VM is across the battery. To check the calibration of the meter, the number scale of 0-80 corresponds with voltages of 8-16 volts, each division above zero representing one tenth of one volt (0.10). For example, 40 on the scale would be 12.0 volts, 41 on the scale would be 12.1 volts and so on. There is no calibrating potentiometer for the meter, but it can be adjusted the same as shown in Figures 18-12 and 18-13.

Calibrating a D.C. Voltmeter

Accurate D.C. voltmeters, for calibration and voltage regulator adjustment purposes are not readily available in suitable ranges at a reasonable cost.

In the absence of an accurate (½% or better) calibration standard voltmeter, a D.C. voltmeter can be calibrated at one point by using a lead-acid battery known to be in good condition. The battery should be fully charged, having a specific gravity of 1.250 or more at room temperature (77 degrees F to be exact), that has not been charged or discharged for at least 24 hours, but not more than 48 hours. The voltage should be 2.1 volts per cell. Table 18-3 shows the voltage for common nominal battery voltages.

Nominal Battery Voltage	Number Cells	Voltage at 2.1 V/Cell	Nominal Battery Voltage	Number Cells	Voltage at 2.1 V/Cell
6	3	6.3	24	12	25.2
12	6	12.6	30	15	31.5
18	9	18.9	36	18	37.8

Table 18-3

The larger and more costly test meters and shop calibration meters, needing repair or calibration, can be sent to a qualified repair and calibration shop for this service.

19
MISCELLANEOUS COMPONENTS

Cables and Clamps

Two kinds of power cables are used in battery chargers. Line cords or primary cables, and D.C. battery or secondary cables.

Line cords often have only two prongs on the plug, and two conductors, one black and one white, or they may have a "U" grounded plug having a third prong, and a third conductor colored green. The green lead or ground wire is connected to the round or "U" shaped prong, and the other end of the wire is connected to the metal frame of the charger. With both types, the size of the conductors is determined by the maximum primary amperage of the charger transformer and other AC loads. Usually, this current is indicated on the charger nameplate. Always use a larger cord than one that is too small. Long cords, or extension cords, especially small ones, will reduce the D.C. output current of the charger. Often that is why a charger will not charge at full charge. From Ohm's law, the power output drops as the square of the voltage, not just in proportion to the voltage. For example, if the voltage drops to 90% of 115, or to 103.5 volts at the charger, the power output will drop to 90% of 90%, or 81% of normal output.

Always feel the plug during test to see if it is the slightest bit warm or hot. A warm or hot plug has a high resistance or a poor connection. Replace the plug, if in doubt. Also, feel the cord where it enters the charger case or strain relief. If it is warm, cut the cord, and re-install it. Even if the cord is not warm, and the charger is old, remove the strain relief, move the cord 2 or 3 inches into the charger and replace the strain relief at a new flexing point, avoiding an early failure. Examine the cord for cuts or worn spots and feel the full length of the cord for any warm spots. Examine the connections inside the charger case, and look for discolored, or overheated joints.

The third wire grounded connection is to prevent shock hazards. If the 115 VAC wall receptacle has the third "U" shaped connection grounded or connected to the white ground wire of the power supply, any contact of the 115 volt circuits inside the charger to the metal case will blow a fuse, rather than shock the operator.

Secondary, or battery, connecting cables, must be heavier than primary cables, or line cords, because of the heavier currents carried.

Wire and Cable Sizes for Various Currents

Amperes	Copper Wire Size	Amperes	Copper Wire Size
3	20 or 18	30	10
6	18 or 16	40	8
10	16 or 14	50	8 or 6
15	14 or 12	60	6
20	12 or 10	80	6 or 4
25	10	100	4

Table 19-1

Table 19-1 shows the correct cable size to use for various currents. These sizes are for copper conductors; for aluminum conductors, use the next larger size. On dual-voltage, or multiple rate chargers, wire size is determined by the highest current involved. Always use the same wire size, or larger, than originally supplied.

If the battery clamp, or cable, gets warm where the cable is connected to the clamp, it indicates a frayed cable. The cable should be cut back 2 or 3 inches and resoldered to the clamp, using only rosin core solder. Or use a new crimp terminal on the cable. If a crimp terminal is not available, make one from a piece of copper tubing that will just fit over the cable. Place this on an anvil, flatten it with a hammer, and drill a hole through the formed lug. Use copper tubing just large enough to fit the cable, by flaring, or swage it with a large pin punch, if necessary. Otherwise, it will be too broad to fit inside the clamp.

On all testers and combination charger-testers, the battery cables have a second smaller wire encased in the same jacket to make it possible to read the voltage at the battery terminals before the current goes through the large cable. If this small wire has an open circuit that cannot be located and repaired, the cable should be replaced. In an emergency, a small wire can be taped to the outside of the large cable.

Figures 19-1a and 19-1b show two common types of battery clamps CL

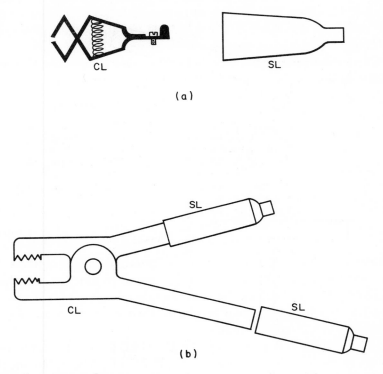

Figure 19-1   Battery clamps

used on battery chargers and battery testers. They may be made of steel, copper-coated steel, zinc-coated steel, or solid copper, and are rated in amperes capacity. Polarity is usually indicated by a plus or minus sign, or by colored sleeves SL (red for positive, and black or green for negative). The clamps shown in Figure 19-1a and smaller sizes of those shown in Figure 19-1b are used for low current chargers. Those shown in Figure 19-1b are used for the high current chargers and testers. A good quality clamp has a copper strap, or cable, connecting the halves of the clamp. If the cable is connected to only one half of the clamp, a poor contact at the battery terminal could send all the current through the spring and rivet to the other half of the clamp and the cable. This overheats the spring so it cannot hold the clamp tightly to the battery terminal. Some clamps have a contact connected to

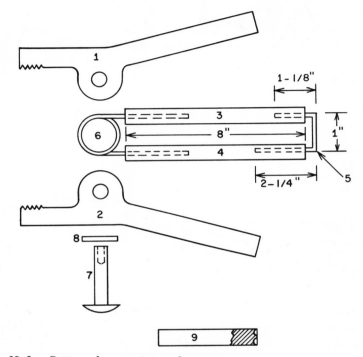

Figure 19-2   Battery clamp spring replacement

the cable end on one half of the clamp, and a pressure contact on the other half of the clamp. Both contacts are insulated from the clamp, preventing spring trouble.

Figure 19-2 shows the tools and procedures for replacing springs on battery clamps if it is economical and the rest of the clamp is in good condition. The clamp insulating sleeves are available separately. Spring repair kits are usually available, and contain a spring pc6, a hollow rivet pc7, and 1 or 2 washers pc8. First, grind off or cut the old rivet on one end and drive it out. Simple tools can be made to quickly replace the springs. Take two pieces of ⅛-inch iron pipe, 8 inches long pc3 and pc4, and insert the tangs of the spring pc6 into one end of each pipe. While holding a pipe in each hand, compress the spring until both pipes can be gripped in one hand, and insert holding piece pc5 (made from ³⁄₁₆″ rod) into the end of each pipe. Then, the clamp can be safely assembled. Be sure the halves of the clamp are meshed so the jaws line up. Insert the rivet pc7 and add washer pc8. Then peen the rivet with the punch pc9, while holding the head of the rivet on an anvil. Rivet punch pc9 can be made on a lathe, or it can be purchased from a hardware, or auto supply store, under the name of "hollow rivet punch" or "brake lining rivet punch."

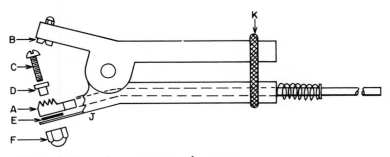

Figure 19–3   Battery clamp contact replacement

Clamps that are corroded or caked with green or white crystals, can be cleaned by soaking for a few minutes in a solution of common household ammonia, either full strength or diluted in the proportion of 1 cup to 1 quart of water. A one pound coffee can that has a tight fitting plastic lid makes an excellent container. Soak the tips of the clamps, and cover with a cloth to prevent obnoxious fumes from escaping. Wash off with water and a wire brush, rub off with a dry cloth, or blow out with compressed air.

Some manufacturers use battery clamps that are insulated from the contact jaws, preventing current from going through the spring and avoiding overheating and loss of spring temper. These contacts are available in kit form, and can be replaced, if the rest of the clamp is in good condition.

Figure 19-3 shows how to replace these contact jaws, using a holding ring K to hold the jaws open safely. This holding ring K can be any round or oval ring having an inside opening of about 2 inches, or one can be formed using a $3/16$ inch steel rod. It should be wrapped with friction tape to prevent slipping.

With the holding ring in position push the cable through the clamp, and install contact jaw A by crimping and soldering. Then, install the contact as shown using bolt C, nylon sleeve D, insulating washer E and nut F to hold the contact A in place on clamp jaw J. Contact B is installed in the same manner. Even with new clamps, this procedure is followed because contact A and hardware C, D, E, and F are supplied loose. However, contact B is riveted in place at the factory.

Heyco Type Strain Reliefs

Heyco Type	A	B	C
		Dimensions	
5P-4	1/2"	7/16"	3/16×1/4"
6P-1	5/8"	17/32"	5/16"
6P3-4	5/8"	17/32"	3/8"
7P	7/8"	3/4"	7/16"
7P-2	3/4"	21/32"	7/16"
8P	7/8"	3/4"	9/16"
8P-1	7/8"	27/32"	5/8"
9P-1	1-1/16"	1"	11/16"
34-1	7/8"	27/32"	5/16×5/8"

Table 19-1A

All cables entering the charger case should have some means of physically holding them secure. The most common strain relief is the two-piece plastic Heyco type having the size number imprinted, such as 8P-1, 9P-1 and so on. To remove or replace these strain reliefs, it is necessary to compress the small insert. Ordinary adjustable "gas pliers" can be used, but they will do a better and quicker job if they are modified as follows: grind out the serrations just inside each jaw, leaving a sharp "eagle beak" tip at the end of each jaw. This sharp nosed tip gets a good bite, and applies pressure next to the metal charger case where it is needed. The jaws should not be over $1/4$ inch thick for use on the smaller Heyco strain reliefs.

The most common Heyco type strain reliefs used in battery chargers and testers are listed in Table 19-1A. Dimension A is the mounting hole diameter, B is the distance across the flats, and C is the bore, or cable size. Those having two dimensions under C, are for flat, or duplex cables. The others are for round cables.

Lamps and Bulbs

Lamps and bulbs are used in many ways in battery chargers, some more critical than others, and require an exact replacement. In pilot

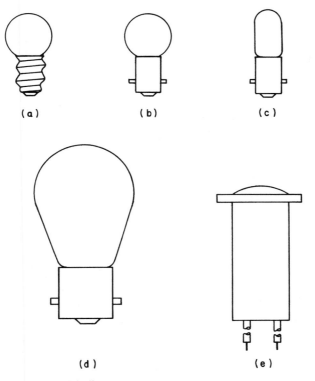

Figure 19–4   Lamps and bulbs

light service, exact replacement is not so critical if the pilot light voltage comes from the transformer or line voltage. Pilot lights operating from voltage from other circuits, such as alternator protectors, voltage regulators and the like, may require an exact replacement.

Most lamps and bulbs are replaceable in the usual manner, but some may be soldered in place, or consist of an enclosed holder and lens and lead assembly.

Lamps and bulbs of the removable type are identified by a number printed on the base, and are standard, regardless of the bulb manufacturer. Some bulbs are standard automotive types and can be purchased locally. Others can be ordered from electronic supply houses or from the battery charger manufacturer.

Lamps and Bulbs Specifications

Mfg. Type	Volts	Amps.	Base	Bulb Style	Fig. 19-4
47	6.3	.15	Bayonet miniature SC	T-3¼	c
51	7.5	1 CP	Bayonet miniature SC	G-3½	b
53	14.4	.10	Screw SC	G-3½	a
55	7.0	2 CP	Bayonet miniature SC	G-4½	b
57	14.0	2 CP	Bayonet miniature SC	G-4½	b
222	2.2	.25	Screw SC	TL-3	a
305	12-18	.50	Bayonet single contact (SC)		d
313	28	.17	Bayonet miniature SC	T-3½	c
433	18	.25	Bayonet miniature SC	G-4½	b
1047	12		Bayonet SC		d
1073	12	32 CP	Bayonet SC		d
1445	18	.15	Bayonet SC	G-3½	b
1815	12-16	.20	Bayonet SC	T-3¼	c
Neon or filament type pilot light or indicator					e

Table 19-2

Electronic catalogs list most of these bulbs giving the voltage, amperage, kind of base, and bulb style or shape. Those most commonly found in battery chargers are shown in Figure 19-4 and Table 19-2. Some lamps and bulbs are rated in amperes, and others in candle power CP. A rough approximation of the amperes drawn can be calculated when only the candle power CP is known. Approximately 1-1.5 watts (use 1.25 as an average value) input is equal to one candle power. A precise figure cannot be determined because of the differences in efficiency of the various constructions used, and the voltage at which they are operating. For example, a #55 bulb is rated at 7 volts and 2 CP.

$$\text{Amperes} = \frac{\text{Watts}}{\text{Voltage}} = \frac{2\,\text{CP} \times 1.25}{7} = .357 \text{ ampere}$$

or 357 milliamperes, which may be between 300 and 400 MA.

If the filament of a bulb is out of place in the bulb, or the filament is sagging, even though it shows continuity, the bulb should be replaced.

Neon bulbs draw practically no current and are frequently used for pilot and indicator lights at 115 VAC.

Knobs and Dials

Knobs and dials are often lost or broken, so they are a common replacement item. The shaft size and shape is the most important consideration in selecting a replacement. Other considerations are the markings on the knob, or if plain, the markings and position of the markings on the panel. Only a few are standard ¼" bore, screwmounted pointer-type dials. Most are peculiar to each manufacturer, and should be ordered from the manufacturer of the charger.

The most common shaft sizes and shapes are: ¼-inch round shaft with a flat for set screw, either slotted or Allen head; ¼-inch round shaft with a flat for spring held knob; double flat with a spring held knob; ⅛-inch square with snap-ring; and small flat shafts such as used on tester switches having a small spring holding knob. Knobs without a set screw are pushed on and pulled off. If they cannot be pulled off easily with the fingers, pry evenly with a broad tool on both sides of the knob to avoid breaking it. On some older, and larger, selector switches, the knob and shaft are one piece and can be removed only by loosening a set screw or cotter pin on the contact arm from inside the charger case.

Some manufacturers can furnish panel decals or labels for renewing old panels with all of the dial markings. In the absence of this, use a label making machine to make new markings and stick them in place.

Wheels and Casters

Wheels and casters on roll-about portable chargers require replacement occasionally. They should be oiled every time a charger is serviced.

Wheels in common use vary in size, having a bore of ⅜ inch, ⅞₆ inch or ½ inch, and an outside diameter from 4-8 inches.

Wheels are usually fastened on the shaft by a snap-cap that is driven on. They can be removed with a special tool, or by following a simple procedure using a punch and a hammer.

Locate the two "dogs" A shown in Figure 19-5a on the snap-cap C that holds the cap tightly on the shaft, and knock them inward slightly

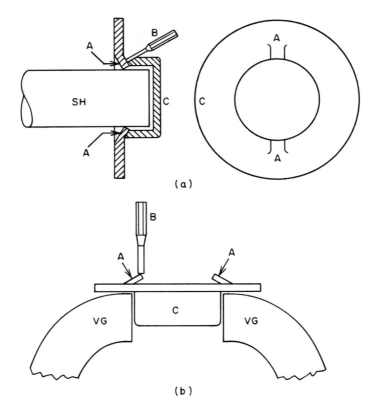

Figure 19–5   Wheel snap-cap removal and restoration

with a small pin punch B until the cap C can be twisted and pulled off with pliers. These can be reused by driving the dogs back into position. This can be done as shown in Figure 19-5b, by laying the cap C between vise jaws VG, or in the end of a pipe of correct size, and drive the dogs A back to their original position with punch B, using the vise jaws as an anvil. The cap C should be a drive fit. Some manufacturers furnish a new snap-cap with each wheel, or they can be ordered separately.

Conversion Tables

Often, the temperature ratings of electrical equipment and electronic devices are given in degrees Centigrade or Deg. C. To convert the

Centigrade temperature reading to the more familiar and common Fahrenheit temperature reading, the following formulas are used.

$$\text{Degrees Fahrenheit} = \frac{9 \times \text{Deg. C}}{5} + 32$$

$$\text{Degrees Centigrade} = \frac{5 \times (\text{Deg. F} - 32)}{9}$$

Some of the most common temperatures encountered are shown in Table 19-3 showing the Centigrade and Fahrenheit equivalents.

Centigrade and Fahrenheit Temperature Conversion Values

Deg. C	Deg. F	Deg. C	Deg. F	Deg. C	Deg. F
20	68	32	89.6	44	111.2
21	69.8	33	91.4	45	113.0
22	71.6	34	93.2	46	114.8
23	73.4	35	95.0	47	116.6
24	75.2	36	96.8	48	118.4
25	77.0	37	98.6	49	120.2
26	78.8	38	100.4	50	122.0
27	80.6	39	102.2	51	123.8
28	82.4	40	104	52	125.6
29	84.2	41	105.8	53	127.4
30	86	42	107.6	54	129.2
31	87.8	43	109.4	55	131.0
0	32			100	212

Table 19-3

SECTION IV

1
DISASSEMBLY, INSPECTION AND CLEANING

Every battery charger that is serviced should be opened for inspection, cleaning and lubrication where needed, even though the cause of failure has been located outside the case. This procedure may prevent a return or early failure. While the case is open, thoroughly inspect all connections: a discolored connection indicates a loose connection or poor contact; clean out dust accumulations; lubricate the timer cam and fan motor; test and examine the rectifiers; examine the transformer winding for signs of overheating, or charring of insulation; check for burned contacts on any open type relays; check for loose mechanical fasteners, motor mounting, rectifier mounting and the like. While the case is still open, operate the charger and look for any unusual conditions, such as sparking or smoking. If the odor of selenium is detected (this odor will soon become familiar), the rectifiers may have arced over or been overloaded at some time and may need replacing. The charger should not be operated for long periods of time with the case open because the air circulation and cooling may be impaired.

To open the case of some chargers is quite simple and obvious. But other charger cases can present a problem when encountered for the first time.

Small chargers are usually made in two pieces, having all parts mounted on the base and panel section. Some models may have the rectifiers mounted on the removable section. Disconnecting the rectifiers from the charger is necessary to make the entire unit accessible for service.

On larger chargers, one or more panels must be removed to gain access to the interior parts. A back panel, a front panel, or both side panels may be removable. A top panel may be the only means of entry. As a general rule, to remove a panel start at the corners or the outer edges. Screws in the center of the panel should not be loosened because they are usually support parts, unless it is certain that they are holding only the panel.

In some cases, the only access is through the front panel, or top panel, on which all of the controls are mounted. In such cases, it is simpler to unfasten the controls from the panel, rather than disconnect the wiring from the controls.

When removing panels, remove only the screws that back all the way out. If there is a nut on the inside, it should not be removed unless the nut is accessible. Never force, bend, or deface any charger to gain access to working parts. Keep in mind that it was assembled in a particular order, and it should be disassembled in the reverse order. A little study and patience will indicate removable procedure. Usually, special tools are not required to disassemble and service battery chargers. Readily available tools will do the job. Any special purpose tools required, such as are shown in other chapters, can be made in the shop.

2
CHARGER TESTING

Every charger being serviced should be given a routine, step-by-step test procedure that will reveal quickly the cause, or causes, of failure.

a. First, remove the access panel, or panels, of the charger case, and make a visual inspection of all parts for any obvious defects, such as burned rectifiers, transformers, and resistors.

b. With the battery clips insulated from each other and not connected to the battery, plug the AC cord into the input tester shown in Figure 1-1, Section II, and adjust the controls for charging by turning on the timer and line switch. The voltmeter reading on the input tester should be below the line voltage as the rate switch is turned through each step. If the voltmeter reading is zero, there is a short. If the voltmeter reading does not move below the line voltage, there is an open circuit in the primary circuit. If there are no shorts and no opens, then turn the input tester to full line voltage and do step c. If there is a short, first check the rectifiers, or disconnect them. If they are good, proceed to the selector switch, transformer and the like until the shorted part is located. If there is an open circuit, first place a jumper wire across the timer and line switch. If still open, connect a test cord, as shown in Figure 1-1, Section II, to the two line cord connections, to test for an open in the line cord and plug.

c. Connect the battery clamps to the battery, observing correct polarity. If there is a large spark when touching the clamps to the battery, and the rectifier is not shorted, recheck the polarity of the clips. It is not uncommon to find that the cable clamps have been reversed by someone.

Check the polarity with a D.C. voltmeter connected to the clamps with the charger on, but not connected to the battery. The meter positive is the battery charger output positive. A small spark may be caused by the current drawn by polarity protector bulbs, or by a very low reverse current through the rectifier. If there is no sparking, plug in the A.C. cord and proceed to test the unit.

d. If the charger has a fan motor, it should be running. If it runs quietly at full speed, just clean and oil it. If it does not run, check for a binding shaft, bent blades, or an open circuit in the motor coil. A slow running motor may just need to be cleaned and oiled. If the shaft has excessive side-play, or is noisy, the fan motor should be replaced.

e. If the unit does not charge, the rectifier conducts properly and there is a D.C. output voltage across the transformer-rectifier, there is an open circuit between the rectifiers and the batteries. Check the D.C. cables, clamps, solenoid, circuit breakers, ammeter and the like, including all connections. If the charger is equipped with an alternator protector, and the light indicates "reverse polarity", check three things: (1) reversed polarity of the battery clamps; (2) open circuit in the D.C. output circuit, poor clamp contact, open (broken) cable, solenoid contacts burned, ammeter open, circuit breaker open, and the like; (3) burned out base lamp, such as #313, 433, or open base resistor.

f. On protector-equipped chargers, the solenoid may "click", but it may have burned contacts, that can not close the D.C. circuit. With the charger set up to charge, and turned on, place a heavy jumper wire across the two large terminals of the solenoid while observing the ammeter. If the ammeter pointer moves upward any detectable amount, the solenoid should be replaced or repaired. Solenoids can temporarily make contact in the heavy contacts. They may work perfectly in the shop, but when moved or shipped back to the customer, the large copper washer contact inside the solenoid may rotate to a bad spot, and not work every time for the customer. Therefore, replace or repair the solenoid when nothing else is found defective, but the charger does not charge every time it is turned on.

If the solenoid does not operate, as determined by the familiar "click", there are several other possible causes, such as an open solenoid coil, a shorted diode across the coil which will cause the ballast lamp to burn brighter than normal, a defective transistor, or a burned out collector or ballast lamp (or resistor).

g. Feel the A.C. cord and plug and the D.C. cords and clamps for any warm or hot spots. Hot spots indicate a high resistance area caused by a poor connection or partially broken cable, which should be corrected. On chargers showing signs of long hard use, remove the strain reliefs, move the cords 2 or 3 inches into the charger, and reinstall the strain reliefs. This gives a new flexing point for continued good service, and prevents premature failure.

TRANSFORMER 3
TESTING

The transformer is the most costly item in a battery charger, and fortunately, the least likely to fail in normal service. Occasionally, a transformer will fail due to lightning flash-over, corrosive atmosphere, flooding, and accident. The best solution is to replace it with an exact factory replacement part. Rewinding is not practical, not only because of the cost, but also because of the critical requirements. It is difficult to duplicate the winding exactly. In many cases, it is impossible to rewind a transformer due to welded or other construction.

A transformer may be shorted, or defective if it tends to blow fuses. Even with the rectifiers disconnected, the short can be in the rate selector switch, especially the box type of rotary or push-button switch. If the switch should connect one side of the line to two taps at the same time, it causes a short and will blow a fuse.

To test a transformer for shorts, plug it into the "input tester" shown in Figure 1-1, Section II with switch SW2 in the "in" position. Switch SW4 should be in the H position for small chargers and in the L position for larger chargers. If the voltmeter reading goes to zero, it indicates a short. It may be a shorted rectifier, shorted transformer, defective rate selector switch, or shorted fan motor windings, to name the most common causes. But there are many parts that could be shorted. First, check the rectifier with the rectifier and diode tester shown in Figure 1-2. If the rectifiers are shorted, disconnect them completely from the transformer secondary, and test with the "input tester" again. If a short still exists, identify and mark each lead to the rate selector switch using numbers or letters on a small tab of masking tape. Remove all but one of the transformer leads from the selector switch, being sure that none of the leads removed is touching another or any other terminal. Be sure that one line lead goes to one lead on the transformer, the other line lead goes to one contact on the selector switch, and that only one transformer lead goes to one contact on the rate selector switch. Test again. If the short is cleared, replace the selector switch. If the short still exists, test the transformer more thoroughly after checking the charger wiring. It is not uncommon to find a charger which has had the wiring changed or altered.

If the transformer is still in doubt, completely disconnect all windings after marking them for identification, and make up a wiring diagram of the transformer windings. The secondary leads are always the thickest. If there is more than one secondary, determine all the leads associated with each secondary by use of an ohmmeter or continuity tester. The primary leads are thinner and have two or more connections to the primary. One line lead, the common, usually goes to the start, or bottom, of the winding. The other end of the primary winding is on top and has one or more taps going to the rate selector switch. Using an ohmmeter on the lowest scale, read the resistance between all primary leads to check for opens, and determine which two leads are the main primary leads for highest rate of charge. Connect these two leads to the "input tester." If the voltmeter goes to zero with no other leads touching each other, then the transformer is shorted and must be replaced. If it is not shorted, then proceed to check for shorts between the various windings, using the "input tester" with SW2 to "in" position and SW4 to "H" position. A short or ground exists if bulb R1 burns.

Transformers should always be replaced with exact duplicates of the original equipment due to the differences in conductivity and output of the rectifiers and diodes used. If a substitute transformer must be installed, be prepared to replace all rectifiers with exact replacements for which the transformer is designed, especially if the output is either too high or too low.

4
TRACING
CIRCUITS

It is very important to know how to make a schematic wiring diagram by tracing the circuits of battery chargers, testers, and associated control circuits.

Usually it is best to start out at the input of the circuit and end up at the output of the circuit, drawing from left to right. It is easy to make a circuit diagram where only one lead goes to a terminal. But when a terminal has more than one lead, then show on the diagram a short stub lead-off from the terminal for each lead, and mark it with any color, number or other identification, shown on the lead. After following only one lead to the next terminal, and so on, then come back to each terminal that has a stub drawn on it, and complete one circuit at a time until the circuit is completed.

For example, trace the circuit of a common Fox alternator protector, part #158200 Model 120 and 121. First, make a lay-out drawing as the parts appear, as shown in Figure 4-1, and later draw it in schematic form, if desired. Although the heat sink HS is actually on top of the solenoid, it is drawn below the solenoid to simplify the drawing. Close inspection will show that the heat sink HS is insulated from all screws and mounting terminals, but is connected to the shell, or collector C, of the transistor Q1. Also, it is connected to the shell of the #1073 bulb B2 by solder, and to terminal 3 of the terminal board.

Starting at the negative battery terminal (−) BAT of the solenoid RL1, draw a line to the shell of the #433 bulb B1 and mark it "Y" for yellow. From the center connection of the #433 bulb B1 draw a line to terminal board terminal 4, and mark it "G" for green. From terminal 4

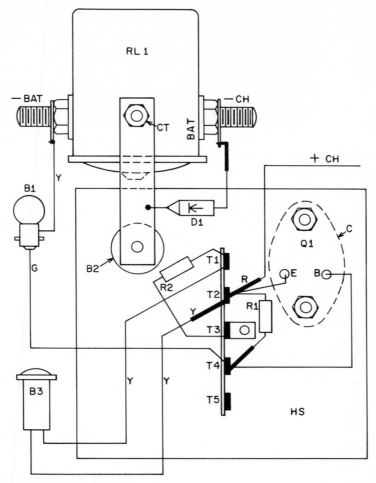

Figure 4–1 Tracing circuits

draw a short stub (all stubs are shown in heavy lines) toward 470 ohm resistor R1, and a line to the transistor Q1, base B. Draw a line from terminal 4 stub to 470-ohm resistor R1. The other end of the resistor R1 goes to terminal 2. From terminal 2, draw two stubs marking one "R" for red and the other "Y" for yellow, and draw a line from terminal 2 to transistor Q1 emitter E. Now, go back to terminal 2 and trace the red "R" stub to the positive output of the charger (+) CH and mark it. Trace the yellow "Y" stub from terminal 2 to the indicator light B3, the other yellow lead from the light B3 goes to terminal 1, through resistor R2 (18 or 47 ohms), and back to terminal 3. The terminal of the solenoid BAT is traced to the negative output of the charger (−) CH, and a stub is drawn out toward the small diode D1. Trace this stub through diode D1 to the copper strip which is connected to and mounted on coil terminal CT and the #1073 bulb B2 which is soldered to the copper strip.

This completes the diagram. It can be redrawn in schematic form if desired and compared with the schematic of Figure 16-4, which is the same alternator protector.

SECTION V

CHARGER CONVERSIONS

Converting 6 Volt Only Chargers to 6/12 Volt Chargers

Many old style, 6 volt only, battery chargers are still serviceable, except possibly the rectifier circuit. But they have little value unless they can be converted to charge both 6 and 12 volt batteries, or at least 12 volt only. With a little study and ingenuity, many chargers can be converted. Most old chargers used a full-wave bridge rectifier of the copper sulphide type, which was less efficient than the newer selenium and silicon rectifiers. A simple conversion from a 6 volt charger into a combination 6, 8, and 12 volt charger can often be done by replacing the old copper sulphide rectifier with an identically rated modern high conductivity selenium rectifier or silicon diodes. The more primary taps or rate adjustments available, the easier it is to bring the charger up to its capacity limits.

Many old chargers using a bridge rectifier will have the secondary wound with two conductors "in hand" and connected in parallel. These are actually two identical secondaries, wound side-by-side that are insulated from each other. By separating these secondaries and connecting them in series with the polarities adding, it is possible to have a full-wave, center-tapped rectifier output saving of half the cost of a bridge rectifier. To double check the connection, the voltage across the two outside leads should be twice the value as that from the center tap to each outside lead. If this was originally a 6 volt only charger, try high conductivity selenium rectifiers or silicon diodes. If sufficient output is not obtained, you must use a bridge rectifier for 12 volts or use the circuit shown in Figure 1-23, Chapter 1, Section I for 6 and 12 volts. With this circuit, the output on full-wave, center-tapped connections is limited to about 85-90% of the output from a bridge rectifier. The transformer primary current, usually stamped on the nameplate, is the limiting factor. The reason for this is explained in Chapter 1, Section I.

For small chargers, a half-wave rectifier can be used if the secondary is made up of two separate windings that can be connected in series. One outside lead is used as the common lead, the center tap is used for 6 volts, and the other outer lead is used for 12 volts. The output of a half-wave rectifier in this circuit should be limited to about half the output of the unit using a full-wave bridge rectifier. Again, the transformer primary current is the limiting factor. This can be limited by using a series resistor in the primary circuit if necessary. This resistor can be a rheostat to adjust the charging current if desired.

Sometimes there is room on the transformer to "snake" through a wire to add a few more turns of wire to the secondary to increase the voltage enough to charge 12 volt batteries.

Another possibility is to use parts of two identical or similarly rated chargers. Use two transformers in one cabinet, if space permits, and connect the transformer primaries in parallel, and the secondaries in series to form a full-wave center tapped circuit.

Converting From Selenium Rectifiers To Silicon Diodes

Selenium rectifiers can be replaced by silicon diodes if high PIV diodes are used. A 300-400 PIV silicon diode has about the same voltage drop and conductivity as a 50 PIV selenium rectifier, especially the newer, high conductivity type.

To make up silicon diode rectifiers, a heat sink is needed. These heat sinks are available without diodes, or they can be made from extruded aluminum stock in the shape of bars, angles and channels. The stock can be purchased in either full lengths or shorter lengths from the scrap pile of any fabricator of aluminum products. They can be cut to length easily with a hack saw. For the press-in type of silicon diode, the material should be at least ⅛ inch thick, or two or more pieces can be bolted securely together before drilling and reaming. For the stud-mounted type of silicon diode, thinner sections may be used. These heat sinks can be mounted on the inside wall of the cabinet to gain extra heat sink capacity. But, they should be insulated from the cabinet by a thin sheet of Mylar insulation or other good heat conducting insulation and held in

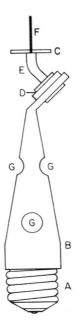

place by nylon screws. Nylon screws are available in size ¼-20 × ½ inch and size of 10-32 × ⅝ inch. Use nylon screws with a steel or brass nut and lock washer, or use two nuts to lock securely.

Converting Bulb Type Chargers To Silicon Diode Type Chargers

The bulb type charger can be converted to use a silicon diode or selenium rectifier in several ways.

Figure 1–1 Silicon diode replacement for tungar bulbs

a. Rectifiers are available that screw into the bulb socket without any alteration, and consist of a finned aluminum heat sink with a press-in type silicon diode having a rating of 12-25 amperes and 300-400 PIV positive base.

b. A copper or aluminum heat sink with a diode can be wired into the charger, eliminating the socket. It can be mounted on the wall of the cabinet and insulated.

c. A simple unit can be made up in the shop from readily available materials that is interchangeable with tungar bulbs. In Figure 1-1, the bulb base A is salvaged from discarded tungar bulbs. All pigtails must be removed to prevent shorting the tip to the shell. Heat sink B is a 5-¼ inch length of 1-½ inch copper tubing, type M, which just fits over the base A after grinding off the lip on the base. If tubing B is cut off with a pipe cutter, it will have to be swaged out to a full 1-½ inch inside diameter. To make a swage, turn down on the lathe a 1 inch iron pipe coupling to a diameter of 1.50 inches, back from the end about ⅝ inch and bevel the end back ⅛ inch. A pipe plug screwed in the other end aids driving with a hammer. Even if the tubing B is sawed off, this swage must hold the round shape of one end of the tubing during the next operation. Drive the swage in one end of the tubing, flatten the other end for 1 inch in a vise, and bend at an angle of about 45 degrees. Reshape the other end to perfectly round by tapping on the tubing held on an anvil, with the swage still in position. Then, drill four holes as shown at G. Drill a hole for the diode $^{31}\!/_{64}$ inch and ream it out to 0.4965 to 0.4985 inch. The same drill can be used for the four ventilating holes at G, or they may be $\frac{7}{16}$ or ½ inch. The tubing B is fitted to base A, after sanding or buffing with a wire brush, and soldered in place with rosin core solder. Apply a good heat sink compound, and press the diode D into the heat sink. The diode lead is bent as shown by holding the lead with long nose pliers close to the diode and bending the end of the lead. Insulator E is a ½ inch length of spaghetti tubing and C is a rubber washer about ⅝ inch diameter cut out of an old inner tube. Prick a pin hole in the center, and slip it over the diode lead to prevent the clip from coming in contact with the diode base. Terminal F is the diode lead that connects to the charger clip, and is the anode connection. These diodes must always be either positive output or positive base to be interchangeable with tungar bulbs.

The silicon diode conversion does not require the reactor as does the tungar bulb. So if the reactor is burned out, wire around it, or remove it. If left in the circuit and it is defective, it may cause erratic operation due to intermittent shorts. If it is in good condition, it does not have to be removed or shorted out. If the charging current is too high on the lowest tap, it may be necessary to add a ½ to 1 ohm resistor in the charging circuit in the same position as the reactor.

The bulb socket of a tungar charger always has the shell of the tube base connected to the D.C. output. The center contact carries only filament current, but the shell, having greater contact area, carries the filament current plus the D.C. charging current. With the silicon diode conversion replacement, only the shell is used, and the filament winding is idle. The charger runs cooler.

Knowing that discarded tube bases just fit the inside of 1-½ inch copper tubing, the ingenious technician can devise many different ways of attaching silicon diodes to this highly efficient copper heat sink. For example, cut a 4 inch length of tubing, solder it to the base, and solder a 1-½ inch copper pipe cap on the other end. Drill and tap a hole in the center of the cap, screw in a stud-mounted silicon diode (12-25 amperes, 300-400 PIV, positive base), and solder a short lead to the terminal if needed for Faehnstock clips. It may be necessary, or advisable, to first braze a nut or thick, solid washer to the inside of the cap before drilling and tapping the hole, and before soldering it in place. Any method used should take into consideration the ease of replacing the diode.

The tungar bulb has a very limited life, and requires frequent replacement. It requires extra power for the filament and for power lost due to the high voltage drop. The silicon diode has none of these objections and should last indefinitely in normal service. If the diode should fail, the entire unit does not have to be replaced since a new diode will renew it again at little expense.

APPENDIX A

Battery Charger Equipment Source Index

Source No.	Manufacturer	Product
1	Alpha Molykote Corp. 65 Harvard Avenue Stamford, Connecticut 06902	*Special lubricants distributed through bearing and machinery supply houses*
2	Allied Electronics 100 N. Western Avenue Chicago, Illinois 60680	*Electronic parts and equipment*

3	Allen Test Products
	Division of Allen Electrical
	   Equipmènt Company
	2101 N. Pitcher Street
	Kalamazoo, Michigan 49007
	   or
	9135 Independence Avenue
	Chatswood, California 91311
		Battery chargers, testers, and automotive tune-up equipment

4	Associated Equipment Corp. 1541 Salzman Street St. Louis, Missouri 63133	*Battery chargers and battery testers*
5	Baldor Manufacturing Company P.O. Box 6238 Fort Smith, Arkansas 72901	*Battery chargers*
6	Berg-Gibson Manufacturing Co. 819-25 E. 17th Street Kansas City, Missouri 64108	*Industrial battery chargers*
7	Big Four Maineville, Ohio 45039	*Battery chargers, automotive test equipment, and parts for Franklin & Litton*

Battery Charger Equipment
Source Index, Cont.

Source No.	Manufacturer	Product
8	Christie Electric Corporation 3410 W. 67th Street Los Angeles, California 90043	Battery chargers and testers
9	Electro-Standards Laboratories, Inc. 3560 N.W. 34th Street Miami, Florida 33142	Parts for all brands of battery chargers
10	ESB Brands, Inc. 2000 E. Ohio Building Cleveland, Ohio 44101	Battery chargers
11	Electric Service Systems 5555 W. 78th Street Minneapolis, Minnesota 55435	Battery chargers
12	Fox Products Company 4720 N. 18th Street Philadelphia, Pa. 19141	Battery chargers and testers
13	Kal-Equip Company 411 Washington Street Otsego, Michigan 49078	Battery chargers, testers, and automotive test equipment
14	Industrial Instrument Works, Inc. 3305 Tchoupitoulas Street New Orleans, Louisiana 70115	Special electric meters. Repairs all types of precision meters
15	Laher Spring and Electric Car Corporation Eastern Division: .P.O. Box 731 New Albany, Mississippi 38652 Western Division: 2615 Magnolia Street Oakland, California 94607	Industrial and golf cart battery chargers
16	Lester Electrical of Nebraska, Inc. 625 W. "A" Street Lincoln, Nebraska 68522 or Lester Equipment Mfg. Co., Inc. 151 W. 17th Street Los Angeles, California 90015	Industrial and golf cart battery chargers
17	Marquette Corporation 5075 Wayzota Boulevard Minneapolis, Minnesota 55416	Battery chargers, testers, and automotive tune-up equipment. Order all battery charger parts from Source 18.

Battery Charger Equipment
Source Index, Cont.

Source No.	Manufacturer	Product
18	Marquette Corporation (formerly Heyer) S. Main and Werner Streets Bangor, Pa. 18013	Subsidiary of Source 17. Supplies products of Source 17.
19	Motor Appliance Corporation (MAC) P. O. Box 22 St. Louis Air Park 22 Mercury Boulevard Chesterfield, Missouri 63017	Industrial and golf cart battery chargers
20	New England Business Service, Inc. Townsend, Massachusetts 01469	Business forms and service tags
21	Pluskota Electric Company Alsip Industrial District 11637 S. Mayfield Avenue Worth, Illinois 60482	Industrial battery chargers
22	Quick-Charge Manufacturing Co. 407 N.E. 48th Street Oklahoma City, Oklahoma 73105	Battery chargers
23	Richardson-Allen Corporation 115-16 15th Avenue College Point, Long Island, New York	Industrial battery chargers
24	Schauer Manufacturing Corp. 4500 Alpine Avenue Cincinnati, Ohio 45242	Battery chargers and automotive test equipment
25	Triple A Specialty Company 5750 W. 51st Street Chicago, Illinois 60638	Silver Beauty battery chargers
26	La Marche Manufacturing Co. 106 Bradrock Drive Des Plaines, Illinois 60018	Marine and industrial battery chargers
27	W. W. Grainger, Inc. General Offices: 5959 W. Howard Street Chicago, Illinois 60648	Fan motors, blades, switches, and various electrical supplies from warehouses in many large cities

APPENDIX B

Battery Charger Brand Names
and Manufacturer Source Numbers

Brand Name	Source Number	Brand Name	Source Number	Brand Name	Source Number
Abel	12	Ford	17	National Accounts	17
Advance	7	Fox	12	Norko	24
Agway	12	Franklin	7	Northern	
Allis Chalmers	19	Gamble Varco	12	Electronics	11
American Battery	12	Goodrich	17	Ordnance	17
American Lincoln	19	Goodyear	12	Pargo	11
Amoco	12	Goodyear		Pluskota	21
Associated		Speedway	7	Quick-Charge	22
Equipment	4	Grant	10	Richardson-Allen	23
Atlas	7, 12, 13, 17	Gulf	12	Riverside	12
Allen	3	Hartman	4	Safetronic	12
Baldor	5	Hester Battery		Schauer	24
Balkamp	12	Mfg. Co.	19	Sears Roebuck	19
Berg-Gibson	6	Heyer	18	Shell	8
Big Four	7	Jordan	10	Silver Beauty	25
Bowers	12	Kal-Equip	13	Spur	12
Christie	8	Laher	15	Starline	18
Cities Service	17	Lester	16	Super	12
Co-op	12	Litton	7	Sunoco	12
Cushman	16	Mac	19	Titan	12
Dresser	11	Marquette	17	Triple A	25
ESB Brands Inc.	10	Mobil Oil	12	Unico	12
Esstron	11	Monarch	11	Union	12
Exide	10	Monitor	6	Union 76	8
Ferromatic	7	Motor Appliance		Viking	24
Firestone	17	Corp.	19	Willard Battery	10
Flying A	12	Mallory	12	World	7
Foremost	17				

APPENDIX C

1. Abbreviations

A	*Ammeter*	MFD	*Microfarads*
AC	*Alternating Current*	NC	*Normally Closed*
B	*Bulb, Black, or Bus-Bar*	Neg.	*Negative*
BAT	*Battery*	NO	*Normally Open*
Bdg.	*Bridge*	NPN	*Negative Positive Negative*
BL	*Blue*	NPNP	*Negative Positive Negative*
Bz	*Buzzer or Horn*		*Positive*
C	*Capacitor, Condenser, Cathode,*	O	*Orange*
	Collector	P	*Primary*
CB	*Circuit Breaker*	PC	*Piece*
CH	*Choke, Charger*	PIV	*Peak Inverse Voltage*
CCW	*Counter Clockwise*	PL	*Pilot Light*
CT	*Center Tap*	PNP	*Positive Negative Positive*
CW	*Clockwise*	PNPN	*Positive Negative Positive*
D	*Diode*		*Negative*
DC	*Direct Current*	Pos.	*Positive*
Deg. C	*Degrees Centigrade*	Q	*Transistor*
Deg. F	*Degrees Fahrenheit*	R	*Resistor, Thermistor, Red*
Diac	*Diode AC Switch*	R-C	*Resistor-Capacitor*
DPDT	*Double Pole Double Throw*	RFC	*Radio Frequency Choke*
DPST	*Double Pole Single Throw*	RL	*Relay, Solenoid*
E	*Voltage*	RLC	*Relay or Solenoid Contacts*
F	*Fuse*	RMS	*Root Mean Square*
FM	*Fan Motor*	RPM	*Revolutions Per Minute*
FW	*Full Wave*	SBS	*Silicon Bilateral Switch*
G	*Green, Generator*	SCR	*Silicon Controlled Rectifier*
HS	*Heat Sink*	SCS	*Silicon Controlled Switch*
HW	*Half Wave*	SE	*Selenium*
Hz	*Hertz-Cycles Per Second*	SH	*Shunt*
I	*Amperes*	SI	*Silicon*
K	*Thousand*	SL	*Slot*
L	*Line, Lead*	SPDT	*Single Pole Double Throw*
M.A.	*Milliamperes*	SPST	*Single Pole Single Throw*
MC	*Magnetic Clutch*	SUS	*Silicon Unilateral Switch*

Abbreviations Cont.

SW	*Switch*	TVS	*Transient Voltage Suppressor*
T	*Transformer*	UJT	*Unijunction Transistor*
TB	*Terminal Board*	V	*Volts, Voltmeter*
TH	*Thermostat, Thermometer*	V.A.	*Volt Amperes*
TI	*Timer*	V.O.M.	*Volt-Ohm-Milliammeter*
TM	*Timer Motor*	VT	*Vacuum Tube*
TMC	*Timer Motor Contacts*	W	*Watts, White*
TMS	*Time Switch*	Y	*Yellow*
TR	*Thermal Relay*	Z	*Zener Diode*
TVR	*Terminal Voltage Relay*		

2. Symbols

AMMETER; ALSO, AMPERE METER AND CURRENT METER – – – – – – – –   (A)

BATTERY, STORAGE: WET OR DRY CELL – =|·|||·||||·||+

CAPACITOR OR CONDENSER –

CIRCUIT BREAKER SWITCH –

COIL OR CHOKE –00000–

DIODE –
SILICON   ZENER

FAN MOTOR WITH BLADES –   (F)

FUSE –

GENERATOR –   (G)

OHMMETER –   (OHM)

POTENTIOMETER –

RESISTOR –

RHEOSTAT –

SWITCH –
SPST   SPDT   DPST   DPDT

TRANSFORMER – – – – – – – – – – – – – – – –
AUTOTRANSFORMER   POTENTIAL   SIDE LINE   POLARITY MARKS
                  TRANSFORMER   CURRENT
                                   TRANSFORMER

TRANSISTOR –   BASE   COLLECTOR
                                                                EMITTER

VOLTMETER –   (V)

RELAY OR SOLENOID –   00000

INDEX

THE AUTHOR

Charles R. Cantonwine E.E. is a graduate electrical engineer who majored in electronics. He has many years of experience in all phases of electrical and mechanical design, manufacturing and servicing. He is a pioneer in police radio, and holds two amateur, and one first-class commercial radio operator's licenses. As an electronics engineer for the Naval Ordnance Laboratory during World War II, he helped develop the famous Holmes anti-torpedo device, and other underwater ordnance. He holds many patents and has others pending. Operating his own electric shop in Hot Springs, Arkansas, Mr. Cantonwine builds special equipment, repairs industrial electronics equipment and motors, and runs an authorized service depot for all major brands of automotive tune-up and test equipment, battery chargers and testors. One of his developments is the Cantonwine Meter, designed especially for the needs of the battery service equipment industry.